Digital China: Big Data and Government Managerial Decision

Qing Jiang
Editor

Digital China: Big Data and Government Managerial Decision

Editor
Qing Jiang
Zhongnan University of Economics
and Law
Wuhan, Hubei, China

ISBN 978-981-19-9717-4   ISBN 978-981-19-9715-0 (eBook)
https://doi.org/10.1007/978-981-19-9715-0

Jointly published with China Renmin University Press
The print edition is not for sale in China (Mainland). Customers from China (Mainland) please order the print book from: China Renmin University Press.

© China Renmin University Press 2023
This work is subject to copyright. All rights are solely and exclusively licensed by the Publisher, whether the whole or part of the material is concerned, specifically the rights of reprinting, reuse of illustrations, recitation, broadcasting, reproduction on microfilms or in any other physical way, and transmission or information storage and retrieval, electronic adaptation, computer software, or by similar or dissimilar methodology now known or hereafter developed.
The use of general descriptive names, registered names, trademarks, service marks, etc. in this publication does not imply, even in the absence of a specific statement, that such names are exempt from the relevant protective laws and regulations and therefore free for general use.
The publishers, the authors, and the editors are safe to assume that the advice and information in this book are believed to be true and accurate at the date of publication. Neither the publishers nor the authors or the editors give a warranty, expressed or implied, with respect to the material contained herein or for any errors or omissions that may have been made. The publishers remain neutral with regard to jurisdictional claims in published maps and institutional affiliations.

B&R Book Program

This Springer imprint is published by the registered company Springer Nature Singapore Pte Ltd.
The registered company address is: 152 Beach Road, #21-01/04 Gateway East, Singapore 189721, Singapore

Dedication

"The key to the application of big data is analyzing and creating value. With big data, government departments can formulate policies scientifically through mining results; Companies can maximize profits; Researchers can discover scientific laws, as well as support social and economic development." This was the point I made at the High-level Forum on Computing and Economic Development in the Context of Big Data, held in January 2013. Given that big data will become an important factor driving social innovation, we suggested at that time three steps for the development of China's big data: The first step is to integrate the government data so that it is no longer segregated; the second step is to create the environment for decision-making based on data, and the third step is to develop the ability of scientific decision-making.

While everyone is talking about big data today, the concept was very new to people ten years ago. The CAS Research Center on Fictitious Economy and Data Science has been conducting research based on data science since 2007 and has also applied the results to national macro decisions, as well as to local governance. There were gains and losses all the way along.

Today, there are still many major challenges facing big data: first, to study the logical relationships between different representations of heterogeneous data to find general rules for "multidimensional data tables" based on heterogeneous data; second, to explore methods for characterizing the complexity and uncertainty of big data and its system modeling; and third, to study the impact of the relationship between data heterogeneity and decision heterogeneity on big data understanding and mining as well as management decisions. However, the environment for the development of big data is improving with more government support and policy incentives. As practitioners, we are pleased and hope that an increasing number of leaders and decision makers will be able to recognize, understand, appreciate, and utilize big data, as developing the right mindset is the basis for making the right decisions

This book, Digital China, does not involve professional knowledge of computing techniques or data analysis, but discusses rather scenarios and events from the author's own experience, such as a vernacular version of data information. I sincerely hope that this meaningful book will bring new values to your life.

Shi Yong
Counselor of the State Council
Academician of the World Academy of Sciences
Director of the CAS Research Center on Fictitious
Economy and Data Science
Director of the Key Laboratory of Big Data Mining and Knowledge
Management, Chinese Academy of Sciences

Special Dedications

Human development is the process of knowing the world and knowing oneself. Human knowledge is constantly expanding, based on which mankind is constantly progressing. If modern human beings are smarter than ancient people, it is solely because of the constant enrichment of their overall understanding. Mankind can only get closer and closer to the truth of things by constantly thinking toward a comprehensive, connected, evolving, and dynamic view. A one-sided, static, isolated view of things is incompatible with evolution. To be wise is to be able to look far ahead, to see things more comprehensively than others, to see things in a connected, evolving way, and to see further, comprehensively, deeply, and thoroughly. The comparison of wisdom is actually determined by who has more comprehensive information and can apply it in a scientific and dialectical way. The cognition and application of big data help people see things more comprehensively.

Digital China aims to help people visualize and understand the application of big data and big data analysis, providing important reference value for looking at issues based on big data. The author, Jiang Qing, is my classmate in the director training course of the National Press and Publication Administration. She has always been a good student and has worked hard. This book is the result of her research and is very useful for our understanding of big data.

We shall have a respectful mindset in the face of big data. Big data can help us make decisions, but it also requires us to conduct an analysis in a comprehensive, connected, and developmental manner. A certain type of data can only reflect a certain aspect of the issue. It must be analyzed comprehensively and combined with specific circumstances and must not be taken over and used mechanically, one-sidedly, individually, or simply. When General Liu Bocheng was leading his army to the Huai River, the last dangerous barrier on the way to the Dabie Mountains,

Dedication vii

the act that he personally measured the depth of the river regarding owning a large amount of intelligence ultimately ensured the smooth progress of the army. Without big data, there would be a lack of basis for analysis, but the mechanical application of big data would only yield mechanical results. Currently, although some data are saved confidentially, it would be useless for people and organizations who do not know how to use the data even if they have accessibility.

As evolves, mankind will come to know more and get even closer to the truth of things. However, the world is ever-changing, and the only constant thing is change. Have you mastered big data yet?

Zhai Xiaochun
Deputy Inspector of the Research Department of the Revolutionary
Committee of the Chinese Kuomintang

In December 2017, General Secretary Xi Jinping pointed out during the second group study session of the Political Bureau of the CPC Central Committee on the implementation of the national big data strategy that it is a basic skill for leaders to perform their job well being good at acquiring, analyzing, and utilizing data. Leaders at all levels should further study, understand big data, utilize big data well, and enhance their skills in applying data to promote various tasks. They should continuously improve their capability to grasp the laws of the development of big data so that big data can play a greater role in solving various tasks.

Countries are paying increasing attention to big data strategies and are trying to lead government decisions and promote social progress through data analysis. China is also rapidly laying out big data strategy, taking the development of big data as an important element and fundamental work in promoting the modernization of the national governance system and governance capacity. For further reformation, utilizing big data well can achieve twice the result with half the effort. The deepening of the overall reform, which extensively involves the analysis of information, is the basic way to grasp the characteristics of the times, historical progress, social development, and economic operation.

In the face of increasingly complicated social development needs, it is no longer enough to rely on experience alone to run large and sophisticated decision-making systems; scientific decisions must be made based on data analysis. At the same time, it is a proper part of modernizing government governance to strengthen human resources on the government side, training a team of civil servants who can utilize and manage big data, and enhancing their ability to apply big data.

Over the past few years, my colleagues and I have received strong support from the authors of this book and their team as we have conducted research on the modernization of national governance systems and capacities. In the process, we have gained a real sense of the irreplaceable methodological value of big data for controlling the development of modern societies.

Sincerely, we hope the research, reflections, and experiences from the author of this book will be useful to readers.

Ding Yuanzhu
Member of the 13th National Committee of the Chinese
People's Political Consultative Conference
Deputy Director of the Department of Society and Culture
of the National School of Administration

In the Internet era, a large amount of data has been generated in the process of business operations, people's consumption, government supervision, and social organization services, which will help us to improve the market order, effectively carry out market supervision, and better serve and protect consumers' rights and interests. It is a concrete action of market supervision departments to practice Xi Jinping's thought of socialism with Chinese characteristics in the new era to actively build a comprehensive data platform for market supervision and consumer protection and strive to form an innovative working pattern of consumer rights protection that combines administrative supervision, social monitoring, and industry regulation in a collaborative manner. In the past few years, we have been working closely with big data in our work on commodity quality regulation and market order maintenance and have been positively exploring and collaborating with the laboratory team of big data led by the authors of this book.

From a lack of understanding of the concept of big data a few years ago to working on specific issues in specific jobs where big data should be applied, big data has become clearer and more useful for us. An increasing number of public servants are realizing the appeal and value of big data and shall quickly pick up knowledge and improve their capability. The publication of this book, Digital China, is timely and will certainly meet the needs of readers well.

Yang Hongcan
Director General, Bureau of Anti-Monopoly and Anti-Unfair
Competition Enforcement, State Administration
for Industry and Commerce

I met the author of Digital China at the inauguration ceremony of the "Big Data Research and Service Base" between the China Statistical Information Service Centre and Xiamen City in June 2014, which is also introduced in this book. I have had the privilege of witnessing the process of applying big data to the ground, and I certainly hope to continually promote the application of big data in local leadership decision-making and economic and social services where it will be truly useful.

Currently, big data are still on the rise and are greatly influencing the value system, knowledge system, and lifestyle of human beings. "In the future, leaders who do not know how to utilize big data for scientific decision-making will be eliminated. Therefore, neither leaders nor normal government personnel cannot ignore the power of data and should strive to develop their capability of thinking and applying big data for decision-making." Digital China is the first book that combines practical examples of applying big data to China's economic and social scenarios. In the book, the author presents her personal experience of the development path, major events, application cases, and changes in the industry development of big data. It is easy to understand from the basic concepts and characteristics of big data to practical interpretations, making it a book that should be kept on the desk about big data.

Xu Wendong
Standing Committee Member, Deputy Secretary
General, and Director of the General Office of the Chinese
People's Political Consultative Conference in Xiamen

With the utilization of computer-based big data and artificial intelligence, people can reproduce, predict, and discover the evolution of the laws and characteristics of the objective world. As a computer scientist, I am very pleased to see that an increasing number of scientists and practitioners in various fields are now joining the research field of artificial intelligence and to see that big data and artificial intelligence can not only improve people's lives, but also influence the shape of society. It is undisputed that there will be many more surprises waiting to be discovered in the future.

This is a golden age for the development of artificial intelligence. The development of big data and artificial intelligence has changed people's thinking and living habits and has driven our society forward. However, the great driving reasons for social progress are, to a large extent, the decision makers in government departments. Only by giving full play to the role of the human brain can the role of big data be maximized and the fundamental purpose of scientific computing in the service of humanity be achieved. This book Digital China is a good introduction to the application of big data, taking the history of its development in China's policy context as the clue, industry-specific examples of the application of the analysis of big data as the content, and developing people's data intelligence as the intention.

Liu Wanquan
Expert in the Beijing "Haiju Project"
Ph.D. in Computer Science, Curtin University, Australia

Utilizing big data to effectively plan the line of the masses in the new era. When I first read the electronic preview version of Digital China, I was studying the important spirit of the second collective learning of the Political Bureau of the Central Committee of the Communist Party of China when XI Jinping, the General Secretary, was presiding over. This book happened to help me further understand some of the key points of the General Secretary's speech. The General Secretary stressed that the development of big data is changing rapidly, and we should review the situation, plan carefully, layout ahead, and strive to take the initiative to gain an in-depth understanding of the current situation and trends of big data development and its impact on economic and social development and to analyze the achievements and problems of the development of big data in China. The introduction of the Industry Development Level Index of Big Data in this book allows us to comprehensively quantify the current state of the development of big data in China, as well as to more deeply understand the urgency and importance of leaders and decision makers improving data capability first to utilize big data to promote the construction of a digital government, promote the protection and improvement of people's livelihood, and safeguard national data security.

The essence of "Big Data + Government" is to take the government service platform as the basis, to make public services inclusive as the main content, and to achieve digital government, smart government, and digital economy as the goal. Years of practice have shown that applying big data to government affairs is the way of the masses in the new era. There is a real need and profound historical significance to utilize big data to link the digital society and the real society, to realize the optimization and reorganization of the government organizational structure and processes, and to build an intensive, efficient, and transparent government governance and operation model, providing society with government management and government service under a new model, new spirit, and new governance structure.

The collection, analysis, research, and application of big data on government affairs is the basis for the government's scientific decision-making, helping to enhance the credibility of policies and the public's understanding and support for policy decisions. Big data on government affairs has become an important way to interview the people, listen to the people, understand public opinion, collect public wisdom, and solve people's worries, promoting the resolution of people's reasonable demands, bringing the relationship between the Party and the people closer, and enhancing the efficiency of the Party and the government's governance. General Secretary XI Jinping has repeatedly stressed the need to make good use of the Internet to walk with the masses. At a deeper level, big data on government affairs has given a new connotation to the mass route and strengthened its core value. The traditional operation path of the mass route is mainly to go to the frontier to understand the people's situation and solve their problems, but in reality, there are still common shortcomings, such as the information flow not being very smooth, which affects the effectiveness. "One who lives in the room knows it when it leaks; one who lives among the folk knows it when the government policy is wrong." Big data on government affairs has greatly innovated the operation mechanism of the mass route, better helping the Party and government to solve various problems encountered in governance and economic construction and development, effectively protecting the rights and interests of the people, and maintaining the close relationship between the Party and the people like blood and fresh.

Digital China can help us to more specifically understand the spirit of the General Secretary's speech on big data and provide us with references for our practical work. Please open this book. I would like to recommend it to everyone to understand the value of big data and learn to apply it through the words in this book.

Liu Liping
Deputy Director of the General Office of the Chinese People's
Political Consultative Conference in Shenzhen

Preface

Main Idea of This Book

As one of the earliest practitioners in the field of big data in China, with the support of successive leaders and experts from the China Statistical Information Service Centre (Social Opinion Research Centre of the National Bureau of Statistics) and the China Economic Monitoring and Analysis Centre of the National Bureau of Statistics, I have unprecedentedly set up and led a team to perform exploratory experiments and practical work on the application of big data in government and enterprises since 2012. Although I was not experienced enough and some ideas are not mature, I have a preliminary understanding of building data-based decision-making based on applications in the process. The application of big data in management work is mainly composed of six steps (WWWHRR): what problem needs to be solved (What); what data are needed to solve the problem (What); where the data comes from (Where); how to get the data (How); and the research application (Research) and the presentation of results (Result) of the data. By solving these problems, it will be easy to apply big data in decision-making.

The foundation of data-based decision-making is data sensitivity. Therefore, I formally introduce the concept of the data intelligence quotient (DQ) in this book, with the aim of promoting the development of data capability in the era of big data. In this case, DQ should be the fourth key element for leaders in the era of big data, following Intelligence Quotient, Emotional Quotient, and Daring Intelligence Quotient. DQ can reflect the differentiation of a person in terms of data decision-making capability. In the future, I will continue to lead my team to further refine the theoretical system of DQ, hoping to provide useful references for the development and measurement of data capability for leaders in the era of big data.

We are engaged in research on the application of big data, with a team culture of trustworthiness, which comes from the original intention of serving leaders and being their management tool. We are grateful for the support and witness of experts along the way, thanks to HUANG Shengming, former Secretary General of the China National Food Industry Association and Director of the Think Tank of the Special

Committee of the National Bureau of Statistics, REN Yuling, former Counsellor of the State Council, WU Zhongze, former member of the Party of the Ministry of Science and Technology, ZHOU Xisheng, Vice President and Executive Deputy Editor-in-Chief of Xinhua News Agency and President of ChinaSo, XU Xianchun, former Deputy Director of the National Bureau of Statistics and Director of the China Economic and Social Data Research Centre of Tsinghua University. ZHENG Yannong, Vice President of China Electronic Commerce Association; SHEN Jing-wang, former Deputy Director of the Department of Information of the State Food and Drug Administration; PAN Fan, former Director of the Research Institute of the National Bureau of Statistics; ZHANG Lingguang, Director of the National Technical Committee 498 for Leisure Standardization; MA Kaili, former President of China Package News Agency; ZHU Wenwu, Deputy Director of the Department of Computer Science of Tsinghua University; HU Jian, President of Xi'an University of Finance and Economics; QI Wanxue, Secretary of the Party Committee of Qufu Normal University; WEN Hao, Vice President of MBA of Zhongnan University of Economics and Law; ZHANG Zhi, Head of Localization of the Graph Database; ZHANG Xing, the famous musician, and many mentors and friends, with your recognition and encouragement, we are able to face the difficulties and make unremitting efforts to continue along the road of the application of big data and contribute to its popularization in China.

Thanks to the trust and perseverance of our partners in the laboratory team of big data and the statistical agency. Some of the contents of this book are also the result of our joint practice. Our efforts will certainly leave a valuable mark in the history of the development of big data in China.

Thanks to the editors of the New Business Knowledge Department of China Renmin University Press who have worked so hard to bring this book to life. Because of your recognition and perseverance, this book, which is written to improve national data capability and the application of big data in China, has been made possible.

Knowledge and experience are all wealth. Having experienced the development of big data in China, there are too many people to list them all, and I will let this book gather my gratitude and blessings.

Everything about big data is about people. Big data is similar to air and water, which is relevant to everyone.

The topics of how to achieve the auxiliary functions of big data for decision-making, improving the DQ of leaders, and applying big data for national governance are of great practical importance. Decision-making based on data analysis is a new and painful change for any organization but one that we must face.

Wuhan, China   Qing Jiang

Contents

Introduction 1
Qing Jiang

The Coming of the Era of Big Data 5
Qing Jiang

Sketch of the Development of Big Data in China 17
Qing Jiang

Managing Changes in the Era of Big Data 51
Qing Jiang

Thoughtless Decision-Making Versus Big Data Thinking 57
Qing Jiang

Leadership in the Era of Big Data 67
Qing Jiang

Enterprise Management in the Era of Big Data 75
Qing Jiang

Big Data Promotes the Construction of Digital China 89
Qing Jiang

Governance Modernization and Governance Digitization 107
Qing Jiang

Data Security and Risk Management 115
Qing Jiang

Big Data, Big Opportunity, and Big Future 123
Qing Jiang

Conclusion—How to Adapt to the Era of Big Data 139

Introduction

Qing Jiang

1 Leaders Should Have a Data-Based View for Decision-Making

The booming Internet has given rise to world-class companies such as Amazon, Google and Facebook, as well as Chinese unicorn companies in the field of the Internet such as Baidu, Tencent and Alibaba. With the rapid development of internet searches, e-commerce, and social networks, a large number of users' personal, settlement, and transaction information will be collected, including the data of search behavior generated by Google and Baidu; the unstructured social information data generated by Facebook, Weibo, and QQ; users' online transaction data generated by Amazon and Taobao; and internet-based e-government information data and statistical data. The rapid growth of data volume provides a solid foundation for the application of big data. These data are a gold mine to be discovered by the government and enterprises, and it is also a value vane for the development of our society behind these data. As big data surges forward, it is only a matter of time before all industries embrace big data, and anyone who refuse may be eliminated.

Big data is both an opportunity and a challenge for wise leaders, who will leave behind traditional decision-making and management styles and embrace the arrival of new management decision-making models in the era of big data. In this era, traditional decision-making has become a key constraint on leaders' ability to make a difference and improve their organizations. Through collecting and analyzing data dynamically in real-time work, government leaders will be able to identify the legacy of past thoughtless decisions, and business managers will identify the barriers to business growth and achieve improvements.

In the future, leaders who do not know how to utilize big data to make scientific decisions will be eliminated. Therefore, both leaders and normal government

---

Q. Jiang (✉)
Zhongnan University of Economics and Law, Wuhan, Hubei, China
e-mail: 18712919290@163.com

© China Renmin University Press 2023
Q. Jiang (ed.), *Digital China: Big Data and Government Managerial Decision*,
https://doi.org/10.1007/978-981-19-9715-0_1

personnel should not ignore the power of data and should strive to develop their own capability in thinking about big data and applying it to make decisions.

2 The Value of Big Data Application

Big data effectively drives the depth and breadth of competition, and at the same time, it is an important factor that constitutes competition. Currently, big data has built the foundation for further applications with unlimited potential. However, most people are not yet ready for those applications, which are still in the early stage. Applying big data to decision making is a strategic priority for organizational development, while insight into data becomes an important sign of core competency for governments and companies in various fields. For internal innovation, government departments can respond to social concerns timelier and accurately through data analysis, providing stronger support for the formulation of policy measures. Through data analysis, enterprises can optimize operational aspects, such as assisting in decision-making, reducing costs, and improving efficiency, to realize delicacy management and interactive marketing.

Big data will bring governments and enterprises new opportunities for scientific decision-making, delicacy management, and service innovation.

For enterprises, with big data, business leaders can understand the work status of each employee and team cooperating situation. Employees can position their roles on the team, improve their own efficiency, and notice the products and opportunities needed by the market. In addition, the value of customer service can be more easily demonstrated for traditional companies by analyzing the data generated in the course of their operations. There is a growing number of companies that take data analytics as their competitive advantage. Many world-renowned companies, such as Google, Walmart, and P&G, are willing to attribute their success to the skillful application of data analytics.

Similarly, for the government, big data can bring delicate social value positioning for government decision-making. With big data, government leaders can understand the daily life of citizens and the development of the city. On the other hand, citizens can enjoy the convenience of travel brought by intelligent transportation, the convenience of medical treatment brought by intelligent medical care, and even the livable environment brought by reasonable economic development. The comprehensiveness and diversity of information for government departments, as well as the rapidity of data generation and the convenience of access, also allow government decision-making to be more rapid and efficient.

At the same time, the significant data growth and diverse data structures of government departments, industries, and enterprises also bring challenges in terms of data processing methods, technologies and application scenarios. The response to adapt to changes in the social environment, using information technology to quickly obtain the most valuable information for one's own benefits from the huge amount of information, will become a key task for economic and social development. Big data

Introduction

can be applied to scan reality, perceive potential problems, warn of possible risks, and predict the development trend of future things. Application to such prediction generally requires multisource heterogeneous data support as well as complex mathematical models and algorithms, while to realize the problem, only a certain aspect of information or simple calculation is needed.

Therefore, the application of big data is not complicated. Its core lies in the correlation and mining of internal and external data, based on which to reveal new knowledge and create new value. The core factor in decision-making is information collection and transmission. In this regard, decision-making and big data are highly compatible. In other words, big data can be seen as an effective management tool for leaders. Therefore, we should make full use of big data from a strategic point of view, flexibly utilizing its value and potential in auxiliary decision-making, to provide better services for organizational development.

3 Utilizing the Voice of Data

The core of the construction of digital China is the modernization and digitization of governance, with the main principle as the scientific and rationalization of government management decisions, which is an inevitable requirement in the era of big data. Leadership decision-making thinking in the era of big data is an important component of governance modernization.

It is of great significance to promote the modernization of the national governance system and capacity by thoroughly analyzing big data and its characteristics, application scenarios, and challenges in decision-making thinking and exploring effective ways to improve decision-making thinking ability by relying on big data.

Today, big data is embedded in all levels of economic and social development and has a profound and comprehensive impact on all fields of our society, especially on the decision-making mindset of the government. As leaders and decision-makers, the basic capacity is to be aware of the spirit of the era in due course, to be digitally literate and to further enhance their overall awareness and sense of relevance to adapt to the changes in the new era. Decision-making requires both full utilization and rational application of big data. Any falsification of data to show off political achievements will cause irreparable damage. In the era of big data, it will become exceptionally difficult to distort or falsify statistical data because the way data are collected, classified, stored, and applied has undergone a qualitative change.

Big data is an upgrade from traditional data and leads to the greater necessity of quantitative analysis as a prerequisite basis for decision-making. The quantitative analysis of data must be combined with in depth, specific research and qualitative analysis, especially at the level of wishes, values, psychology, and spirituality of public opinion. Differences in questionnaire design, research methods, and statistical channels can also result in significant differences in information. Utilizing big data to collect and analyze fragmented information from all aspects of economic life can effectively improve the comprehensiveness and objectivity of information.

Big data comes from all aspects of economic and social life with a wide variety of data sources, making it possible to observe human behavior and collective consciousness more accurately. However, data collection has limitations and cannot be used in a simple and mechanical way. It is important to respect and make good use of the think tank, allowing experts from different fields to collaborate with each other to make more essential, qualitative, systematic, dialectical, and forward-looking analyses and judgments through data. The selection of clear, specific subject data for the target system is the core of decision-making thinking, where clear causal and logical relationships come into play; taking other related data at the periphery, where systemic and interactive relationships come into play. Data influence thinking and vice versa, which is a two-way mechanism of action in between. This will lead to data distortion in the process of data selection, data application, and data interpretation with narrow horizons, biased perspectives, conservative thinking inertia, and rigid ideologies.

The discrepancies between data sources can be caused by factors such as social stratification, the division of labor, and the poverty gap themselves. It is illustrated that there are complex challenges faced in leadership decision-making from phenomena such as data falsification and subjective and deliberate monopolies and distortions of information uncovered by the fight against corruption. Decision-making thinking needs to be alert to data distortions and prevent data traps and data distortions. There can be a gap, sometimes a large gap, between the possession, integration, release of data and its true validity. It is increasingly necessary for decision makers at all levels to pay the necessary attention to certain profound, acute, and sensitive issues and to examine them in a dialectical manner, rather than simply and crudely blocking, deleting, and falsifying data. As long as we naturally allow the huge and complicated data and information generated in the development process to gradually become a deeper component of the economy and society itself, it will be easy for us to know accurately, quickly and without prejudice the "what", and thus to explore the "why". Many cutting-edge, time-sensitive issues can also be touched upon. The wisdom of forward-looking and predictive decision-making can effectively avoid short-term behavior in the pursuit of political success.

Currently, the openness and sharing of government information is an undeniable and obvious effect of big data, with disclosure and feedback complementing each other. On the basis of public government data, the opinion and wisdom provided by the public are increasingly adopted; thus, the various public data can be filtered and verified to build the foundation of social acceptance and recognition of decision-making. This is also reflected in the one-sided requirements in the new era for improving decision-making—the people-centered philosophy.

Utilizing the voice of data well will get us infinitely closer to the truth of things and obtain more scientific results in the era of big data.

The Coming of the Era of Big Data

Qing Jiang

1 What Is Big Data

1.1 Origin of the Concept of Big Data

The concept of big data was first introduced in 1980 by the famous futurist Alvin Toffler in his book "The Third Wave". The concept did not attract much attention and was not widely disseminated until 2011 when McKinsey Global Institute publicly released the study "Big Data: The Next Frontier of Innovation, Competition, and Productivity", which formally stated that the "Era of Big Data" had arrived, that the term "Big Data" began to gain widespread attention. According to the study, "data has permeated every industry and business function today and has become an important factor in production. The mining and utilizing of big data heralds a new wave of productivity growth and consumer surpluses." The concept of big data has only since being popular globally with the promotion of the Hype Cycle by Gartner, and the book "A Revolution that Will Transform How We Live, Work, and Think" by Viktor Mayer-Schönberg and Kenneth Cukier in 2012.

Launched at the World Economic Forum in Switzerland in early 2012, "Big Data, Big Impact" reported that "data has become a new economic asset class, like money and gold." This is in effect a reversal of traditional thinking.

In March of the year, the United States White House Office of Science and Technology Policy released the "Big Data Research and Development Initiative", and organized a high-profile steering group of big data to coordinate and manage more than $200 million in government investments in the field. This means that the United States has elevated big data to a national level and formed a national strategic pattern of mobilization. The Obama administration even defined big data as the "new oil of the future", arguing that the scale of data a country has and its capability to utilize

---

Q. Jiang (✉)
Zhongnan University of Economics and Law, Wuhan, Hubei, China
e-mail: 18712919290@163.com

© China Renmin University Press 2023
Q. Jiang (ed.), *Digital China: Big Data and Government Managerial Decision*,
https://doi.org/10.1007/978-981-19-9715-0_2

it will become an important factor in comprehensive national strength and that the ownership and control of data will become the focus of competition and rivalry between countries and enterprises.

Since then, British Prime Minister David Cameron's new concept of the "right to data" has once again strongly impacted people's habits of thought.

So, what exactly is big data?

The first impression most people have is a lot of data, very large-scale data, and data that is difficult to process. Wikipedia defines it as "Big data is a field that treats ways to analyze, systematically extract information from, or otherwise, deal with data sets that are too large or complex to be dealt with by traditional data processing application software." Big data is characterized by its volume, variety, and velocity and covers a wide range of fields such as the internet, economics, biology, medicine, astronomy, meteorology, and physics, and is a general term for the vast amounts of data and information collected from a variety of sources (such as enterprises, governments, industrial management departments, the internet, email, videos, images, and social media). Ninety percent of the world's data has been generated in the last few years, and the vast majority of valuable data still resides in government and other relevant departments, which is not yet become public available in order.

From data processing in the 1960s, information applications in the 1970s and 1980s, and decision-making support models in the 1990s, to data storage and mining at the beginning of the twenty-first century, the term "Big Data" has not been used until today. Most of the technologies and analytical applications related to big data in China started to emerge around 2010. Hence, big data is still in the early stage of development in China today.

What is big data? In my view, big data is the integrated use of new technological methods to integrate and process multi-source, heterogeneous, and dynamic digital resources at scale, and to integrate information by constituting new, complex logical structures to help people solve specific problems. Big data is an evolution of information technology-based decision-making support systems and can be seen as statistics taking on the wings of information technology.

1.2   Application-Based Interpretation of Big Data

Dr. HUANG Jian, General Manager of Cyphy Technology (Xiamen) Co., Ltd, who was selected as one of the "Thousand Talents Plan" of China, gave his explanation on the application of big data at the first Big Data Forum 2015, "big data should be about not only the quantity but also the correlation between data. Data that were not originally thought to be related, can produce some related results after the analysis, which may or may not have been originally thought of. With supercomputers, some special algorithms are used to find the correlation of the data and also to find out the value of the correlation, which is the application-based interpretation of big data."

It is the application of big data that applies semantic search technology and recommendation algorithms to e-commerce platforms or media platforms to improve user experience. When shopping on e-commerce platforms, you will find that the experience of searching and purchasing is getting better and better, which is the application of semantic search technology. The use of data for text semantic analysis, synonym mining, machine learning, etc. will lead to a significant increase in the transaction rate of online shopping, which leads to increased turnover for merchants. Similarly, the TouTiao is also making full use of recommendation algorithms based on big data to make projections about readers' reading preferences according to their web browsing trajectories and personalize recommended content in real-time to enhance their reading experience, thereby gaining a large number of users, then achieve profitability and development of the business through models such as advertising.

It is the application of big data that analyzes social media data such as Weibo, WeChat, and Twitter to discover user characteristics and provide precise services. Nowadays, many consumer goods companies will carry out data analysis with the intention of precise marketing to increase brand or product loyalty and consumption. And some commercial organizations are often tempted by profit to violate consumers' privacy, which requires special attention.

It is the application of big data that is the real-time pricing mechanism based on the SAS system. This mechanism allows stores, shopping malls, and supermarkets to adjust the prices of tens of thousands of items in real-time, based on customer demand and stock availability, to respond to the market's pricing strategy and maintain a competitive edge.

Even though, in terms of data structure, more than 80% of the massive amounts of data originating from the web and the cloud are unstructured, the current data landscape offers new opportunities to discover and create value, enrich business intelligence, and support leadership decisions making. Of course, there are challenges such as complexity, security, and privacy risks facing big data. Traditional BI (Business Intelligence) can no longer meet the needs of business development. While we are often exposed to several enterprise-level BI platforms, these traditional BI platforms only enable post-reporting and lagging forecasting. It is time to start building models that can truly predict customer loyalty, and use multiple variables to analyze predictions based on historical transaction data to identify customers that are about to be lost or orders that are about to be filled.

In addition, big data has redefined the scope of data management, evolving from data capture, transformation, and loading to new technologies for purifying and organizing unstructured data. The new data management systems are designed to meet the challenges brought by big data, such as distributed database technology, an open-source platform that is currently the most widely used for managing storage and access and processing large data sets at high speed and in parallel. However, distributed database technology is a challenge for many SMEs or government departments, which often do not have the necessary expertise and experience to apply big data and require external resources to help. The application of big data requires skills that are not purely technology-based, but finding the right people with the right skills to analyze big data is the biggest challenge for practical applications. For most organizations, it is difficult and expensive to find and select competent data specialists.[1]

Big Data is mainly derived from local data, Internet data, and Internet of Things (IoT) data. Local data is ubiquitous. Human beings have been recording all kinds of data since the invention of writing. Before the Internet became widespread, most data were stored locally and were not a public data resource. For example, government statistics, citizen consumption data, and enterprise operation data have been deposited over the years in enormous amounts. Once open, it will become a huge treasure trove of data to be mined by researchers. With the popularity of the Internet, people generate billions of Internet data every day when using it. For example, after the emergence of Google Maps and Baidu Maps, a large amount of new location data representing behavior and habits has been generated; with the rise of social media such as Weibo, Facebook, and Twitter, users can share content on the internet anytime and anywhere, thus generating massive amounts of user production data; the boom of e-commerce has brought data on payment behavior, purchasing behavior, logistics and transportation, and more. These massive amounts of internet data hide behaviors and habits that represent specific groups of people, which can help companies accurately identify the factors that influence user behavior and effectively classify customer needs after analysis and mining so that they can be both creative and efficient in meeting the needs. The IoT is an important part of the new generation of information technology and an important development stage of the information age. Its client has been extended and expanded to the exchanging and communication of information between any object and item, so its data volume scale, data generation frequency, data transmission rate, data diversity, and data authenticity are all better than the traditional Internet. The development of big data is inseparable from the IoT, which provides sufficient and favorable data resources for big data, and big data technology also drives the development of the IoT.

2 What Big Data Can Do

At present, the National Bureau of Statistics of China has applied big data to government statistics. The research on the prediction of the real estate price index, Beijing's mobile population monitoring, and the China Statistical Information Service Center's comprehensive evaluation of brand reputation has all achieved good results. In the field of financial investment, data assessment is required when the financial institutions considering whether the credit system of the enterprise is worthy of a loan without the factory plant as collateral before lending small amount of loans for SMEs. Not to mention the social research that there are now many social topics can be fully implemented and realize data collection and research based on data from the Internet or other sources. When big data is applied to management decisions,

---

[1] GAO Changshui, JIANG Daohui, JIANG Qinyun. The Application of Big Data in Government Department [J]. The Technology of IoT, 2014(6): 6–10.

the first to benefit are leaders in government and in all industries. Big data can help leader to manage more effectively, giving every leader the advantage of innovative management and decision-making models, thus to make more scientific decisions.

Big data is an ecosystem that promotes the full integration of society and cannot be understood simply as just data or industry, as all aspects of the economy and society intersect with big data.

One of the application scenarios is dynamic marketing campaigns for specific users through predictive models. The European gaming industry and the Hong Kong Jockey Club can use software to analyze billions of transactions as well as customer characteristics to guide bets and predict outcomes. Haier Group has entered into a partnership with Big Data Research Laboratory of the China Statistical Information Service Center (Home Big Data) to build a Big Data Lab for smart home appliances. Through its after-sales service platform and quality complaint platform, it collects all user complaints and feedback data on Haier products, then analyzes and mines internal and external data to conduct more comprehensive data monitoring of its range of products and conducts proactive maintenance to reduce overall energy consumption, improve user satisfaction and enhance brand reputation. The company is also able to better realize the purpose of the development on service by analyzing data to discover users' preferences, consumption, habits and consumption power, thus guiding the company to carry out dynamic customer maintenance and target marketing.

Another application scenario is the analysis of crime data based on seismic prediction algorithms as a way of predicting the probability of crime. In areas where the algorithm has been applied in Los Angeles, United States, the probability of burglary and violent crime has dropped significantly. In China, Home Big Data has experimented with predictive algorithms to analyze perception data affecting social safety as a way of predicting social risk prevention and assessing the level of security perception in each region. The Sichuan Social Opinion Research Center has applied this type of data set to a local perception of safety survey study, making the public's perception more specific and precise, as well as giving clearer clues to social risk management, which has yielded good results.

Big data can also automatically adjust waiting times or display content through video analysis. Traffic congestion in megacities such as Beijing and Shanghai has seriously affected people's travel and urban living experience. Traffic management departments can analyze the length of the pedestrian queue and vehicle length waiting to cross the street at intersections through video, and then automatically change the waiting time for traffic lights to ease traffic. If no one is crossing the street and the car is waiting for a red light, it can automatically switch to a green light to ensure smooth vehicle passage and reduce vehicle congestion. In addition, fast food restaurants can recognize and analyze the length of queues based on the number of people ordering food in order to automatically display the contents of the electronic menu. If the queue is long, food that can be served quickly is displayed; if the queue is short, food that is more profitable but has a relatively long preparation time is displayed.[2]

---

[2] ZHAO Enuo. Finding New Paths of Cultural Communication [N]. People's Daily, 2014-08-21.

The China Electronic Port integrates data from a dozen departments including industry and commerce, taxation, customs, foreign trade, foreign exchange, banking, public security, transportation, railways, civil aviation, and state inspection. The real-time networking and analysis of the data allow the Electronic Port faster in customs clearance and more efficient in combating law-breakers while becoming a "weather forecasting station" for China's economy, providing very fine and comprehensive decision-making support for national macroeconomic regulation.[3]

The Internet gave birth to the era of information technology, and information technology made the era of big data. Facebook, Twitter, Weibo, and other social media have been introduced one after another, opening up a new era of the Internet. Before that, the main role of the Internet was to disseminate and share information, and its main form of organization was the website, which is static. After entering the new era of the Internet, the Internet began to become a vehicle for real-time interactive communication. On August 23, 2011, a 5.9 magnitude earthquake hit Virginia, and New York City residents first saw the news on Twitter, and only seconds later did they feel the waves coming. Social media has brought the speed of human information dissemination into an era when it is faster than seismic waves.[4]

Nowadays, browsers are also long gone from being the only way to access the Internet. It was relying on computers or laptops to access the Internet, but now almost everyone uses the smartphone. Smart bracelets, smart glasses, and all sorts of smart electronic devices such as GPS navigation equipment have also become media for accessing the internet. In turn, these accesses have become the ways and source of data generation, through which everyone's data is generated and collected.

With the development of information technology, new concepts such as mobile internet, social networks, cloud computing, and blockchain have entered people's daily work and lives one after another, along with the amount of data worldwide exploded. Data has become a fundamental element in the production and operation of enterprises. Analyzing user behavior through big data helps enterprises to get closer to consumers in terms of technology and innovative business models, thus improving business operations and enhancing efficiency, which will be the core competitiveness of modern enterprises. Today, massive amounts of data are being generated all over the world all the time without stopping. The massive amount of data is constantly being collected, exchanged, managed, analyzed, and integrated, becoming an important part of the global economy and society. And among these enterprises are the biggest direct beneficiaries of the application of big data to generate value. In the era of big data, companies want to stand in the competition to be invincible in competition in the era of big data, enterprises must not only own more qualified data but also have sufficient decision-making leadership and scientific management system.

The accumulated value of big data has triggered a profound change in business and management models, and people must adapt to the situation facing this challenge of "Big Data".

---

[3] DU Xiaoyong, FENG Qina. Three Key Concepts of "Data Governance"—from Internet Thinking to Future Governance Picture [J]. People's Forum—Academic Frontiers, 2015, (1) Volume 2: 49–61.

[4] XU Zipei. Big Data and Its Causes [J]. Science and Society, 2014, 4(1): 14–26.

3 Data Acquisition and Application

3.1 The Shape of Big Data

Applying big data should start with applying the small data, which requires developing a sensitivity to data. Big data evolves gradually from small data, which is a normal development ecology. For organizations to turn big data into a competitive advantage, they must build their own data systems. First of all, it is necessary to find the core data internally. The core data is now for many companies actually their own user system, which is the most important. Secondly, find relevant external data and collect it. Again, the data should be acquired from conventional channels. Finally, acquire external social or unstructured data, what is now called social media data. If we look at data forms, we can also take business data, interaction data, and IoT data to help us understand big data.

Business Data. Business data refers to data generated from business channels such as ERP systems, POS terminals and online payment systems, and is now the primary and most traditional source of data for the applications of the business big data. In terms of online e-commerce platforms, they track a large amount of consumers' information such as purchase records, browsing histories, time on page, choice of shipping method, user marketing feedback, and personal details. By collecting, organizing and analyzing this data, e-commerce platforms such as Amazon, Jingdong and Alibaba can optimize their product mix, carry out precision marketing and fast shipping method. Walmart has already marked every product in data in the traditional retail chain, and with consumers' purchase lists, spending amounts, purchase dates, weather and temperature on the day of purchase can also be data in the system. By analyzing unstructured data such as consumer shopping behavior, product correlations can be discovered and product displays can be optimized; by analyzing consumer shopping data, beer and nappy products, which are not related to each other, can be bundled and sold. Walmart not only collects these traditional business data, but also extends its data collection to social networks. When customers talk about certain products or express certain preferences on Facebook or Twitter, such data is recorded and used by Walmart. Today, Walmart is able to optimize the product mix of supermarkets in a consumer's area based on what they say on social media, and can also help consumers mark where the products they are talking about are located in the supermarket.

Interaction Data. Data in the Internet is extremely mixed and most are difficult to exploit. Much of what is recorded by social network data is what users are doing, thinking, and interested in, along with their gender, age, address, occupation, educational background, interests, and etc. Larry Page, the former CEO of Google, is known as the world's greatest data scientist. Google has been collecting data and analyzing to build product mix for over a decade. Google's viewfinder cars would run around the world with panoramic cameras to collect street views of the vast majority of the world's cities; its 3D infrared cameras completed the scanning of tens of millions of books… Through the continuous collection and utilization of data, Google's search, translation, advertising, music, and other products are supported by massive amounts of data and have received reckon positive feedback from users. In addition, Google collects the words that users mistype when searches, stores these errors and links them to the final correct input and use in the development of Google AutoCorrect and Google Translate. The analysis and application of users' behavioral trajectories have also made Google a veritable big data application company.

Facebook, Weibo and WeChat already store personal information shared by billions of users, such as gender, age, and interests, while personal life timeline records the stories of individuals' lives in the pass. With a huge amount of data acquired through personal information and timeline trajectories, these social tools are like the user's memory, remembering the past and present clearly and predicting the user's future. At the same time, we find that the use of these social tools is free for users, but businesses can also reach potential target customers through analysis and thus achieve precise and effective advertising, which has become one of their effective ways to gain benefits. Moreover, the more data users leave behind when using these applications, the better these companies will know their users and the more accurate their advertising will be, which is the usual pattern of the "fleece comes from the pig" business story we often hear.

IoT Data. The most imaginative way for data collection in the future will be IoT data consisting of clusters of sensors. Sensors are already widely distributed in many places, such as offices, houses, transportation systems, factories and supply chains, smart phones, wearable electronic devices, and etc. It is possible to track the location of objects data such as placement, heat, amplitude, pressure, and sound. For example, by placing sensors on cattle in the pasture, the status of each individual cattle in the herd can be monitored at all times. When a cattle is sick or pregnant, the sensor sends the information to the owner who can then take appropriate action in time. The growth in the number of sensors and their ultimate capacity is staggering. The amount of data generated by sensors will overtake social networks as the second largest data source.

3.2 The Applied Object of Big Data

As shown in Fig. 1, Big Data Research Laboratory of the China Statistical Information Service Center (Home Big Data) concludes that in China, the applications of big data are mainly reflected in three objects: enterprises, governments, and research institutions. Through these three objects, the integration value of big data will cover all aspects of economic and social life.

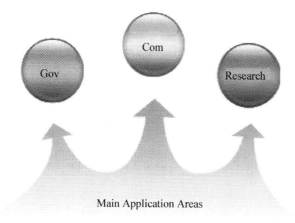

Fig. 1 Domestic big data application areas

3.3 Enterprises: Based on the Aim of Maximizing Profits

At present, the total amount of data resources in China is insufficient, which is more due to the closed nature of our enterprises in terms of data. Relevant surveys show that more than half of the enterprises in China have more than 200 TB of data, most have between 50 to 500 TB, and 1/3 have been using a lot of external data and Internet data. With the change of enterprise management thinking and the reformation of enterprise management and operation mode brought by big data, the total amount and quality of data resources of Chinese enterprises will be further improved, and the application of big data in enterprises will become increasingly important. The main channel for enterprises to maintain or develop their competitive advantage and create value that satisfies consumers is to make profits by providing products and services.

3.3.1 Precise Marketing to Meet User Needs

The key point of precision marketing focuses on how to apply big data to better understand users and their preferences and behaviors. To gain a more comprehensive understanding of customers, companies are keen to collect social data, browser logs, review text, and sensor data to create data models to analyze and predict on demand. A more familiar case in point is the US retailer Target's use of big data analysis to accurately predict at what point customers want to have children. Similarly, through the utilization of big data, telecommunication companies, banks, and insurance companies can better predict the customers will leave, consumer goods companies such as Walmart can more accurately predict which products will get big sales, car manufacturers and the insurance industry will understand the needs and driving levels of their customers, and governments will be able to understand the preferences and real thoughts of the public.

3.3.2 Business Process Optimization

We can dig out valuable information through the analysis of social media data, network search data, and behavioral trajectories to help optimize business processes. One of the widespread applications is the tracking of freight vehicles based on the identification of the geographic location and radio frequency, as well as the use of real-time data on traffic routes to develop better routes to achieve optimization of supply chains and distribution routes. At present, basically, all cars and trucks on China's highways have real-time road guidance information, and Baidu maps and Amap are frequently applied by drivers. The Human Resource (HR) business is also a good application area for big data. HR of companies or HR-like websites can optimize the recruitment process through big data analysis and precisely match CVs with jobs.

3.3.3 Smart Financial Trading

New media is characterized by fast and timely communication and a large reach, whereas financial trading is a major battleground for the application of big data, with many equity trades using big data algorithms to make trading decisions. With the rapid growth of the internet, the influence of social media and website news is now increasingly being considered, when people predict and decide whether to buy or sell using suitable algorithms.

3.4 Government: Based on the Aim of Improving and Upgrading Public Services

Big data governance is an important area of the application of big data, and the position of big data in national and social governance is becoming increasingly important. It has become a "five-dimensional government" with data as the core dimension of social governance, superimposed on the three-dimensional space and one-dimensional time. Internet platform enterprises utilize use the powerful technical capabilities of cloud computing and big data to gather massive amounts of data far beyond the scale of traditional enterprises, complementing the government's historical and authoritative statistics for each other. The advantages of "Internet

The Coming of the Era of Big Data

application + cloud data platform + government governance" will become a representative model of shared governance in the future society.[5] The government is the executive organization that maintains national stability, ensures the basic rights of citizens, achieves sustainable development, and promotes economic growth. Big data is a powerful force for economic growth, transformation, and upgrading, giving rise to new changes in government management and public governance. With the advantages of abundant data resources and application markets, as well as the most promising market application capacity in the world, China has the foundation to become a country with a large amount of qualified data, thus it is an important task at present to make a breakthrough to enhance government governance with the application of big data as soon as possible.

3.4.1 Social Governance, Integrity Building

Big data is now widely used in social governance, particularly in security and law enforcement. The most common application of big data within the public security departments is to assist in the apprehension of criminals, while most commercial enterprises can apply big data methods and technologies to assist in their operations and in response to cyber-attacks The banking and security industries have long applied big data as an important tool for credit collection.

3.4.2 Smart City, Smart Life

Smart cities have become the pursuit of new cities. Many cities in China are engaged in the construction of smart cities as well as analysis and pilot projects of big data. Based on real-time urban traffic data and the use of social network data and weather data, the latest traffic situation is optimized and such intelligent transportation has allowed modern cities to effectively alleviate traffic jams; weather forecasts and GPS navigation based on satellite data make our lives more convenient... Big data that is no longer mysterious is increasingly being applied to people's daily lives. It must be data-based that various cities trying to realize intelligent life for citizens, and gradually intelligent transportation and medical care afterward.

3.4.3 Legislative Support, Assist in Decision-Making

In the process of building digital China, exploring the potential value of big data and researches on massive amounts of information using cloud computing and other technologies can provide scientific support for relevant decision-making. For example, a comprehensive analysis of "information dynamics + sampling information +

---

[5] GAO Hongbing. On the "Five-Dimensional Government" in the Era of Big Data [N]. China Industrial and Commercial News, 2015-07-21.

product information" on infant milk powder can help departmental leaders dynamically grasp the regulation of the industry, identify problems, and decide whether to continue the focus on regulation. If the data is sufficient, it is also possible to clearly analyze the current situation of the industry in a certain region, comparing it with the whole country horizontally and with itself vertically, allowing the development of the industry more objective, which is conducive to regional development without duplication of construction and investment.

3.5 Research Institutions: The Aim of Promoting the Application of Results Through Data Quantification

Unlike Europe and the United States where big data is being rapidly applied to various industries, the main role of big data research in China is currently in economic research, medical care, environmental protection, and education research, while China has also accelerated the pace of turning research into practice. Most of traditional research institutions are knowledgeable in theory, but it is embarrassing that many researchers, including professors in universities and experts in research institutions, presented their work in the form of papers or books of research results they worked hard to achieve. The reason that there are few opportunities to participate in practical scenarios during the research process, is the results are either impossible to realize or many problems occurred in the process. What can help researchers in the era of big data is that, on the basis that everything is quantifiable, researchers can join hands with data agencies to validate their research results in practice based on practical application scenarios and increase the value of their research results.

Sketch of the Development of Big Data in China

Qing Jiang

1 Evolution of Big Data in China

On 27 December 2012, the Director of the National Bureau of Statistics MA Jiantang shared his opinion of the 'Era of Big Data' at the National Conference on Statistics and made clear requirements on how government statistical departments should respond to the opportunities and challenges along with them. This was the first official announcement of facing the era of big data in the country.

On 30 December 2012, the President of the Chinese Academy of Sciences, Academician BAI Chunli called out at the China Sciences and Humanities Forum that a national big data strategy should be developed in China. He proposed the contents of the strategy which should include building a big data research platform to achieve breakthroughs in key technologies; building a benign ecological environment where supportive policies, industry alliances, and industry standards are formed; building a big data industry chain to promote the effective linkage between the innovation and industry.

On 19 January 2013, in response to the appeal of Academician BAI Chunli, more than 40 domestic and oversea scholars, including academicians, representatives of government departments and enterprises who have practical experience, together discussed the value of the application of big data at the High-level Forum on Computer and Economic Development in the Context of Big Data held at the CAS Research Center on Factitious Economy and Data Science which was initiated by Professor CHENG Siwei (deceased) and Academician LI Guojie. All participants unanimously appealed for involving the development of big data in the national strategy. According to the organizer of the forum Prof. SHI Yong, the application of big data is to analyze and create value. With big data, government departments can use the analyzing results

Q. Jiang (✉)
Zhongnan University of Economics and Law, Wuhan, Hubei, China
e-mail: 18712919290@163.com

© China Renmin University Press 2023
Q. Jiang (ed.), *Digital China: Big Data and Government Managerial Decision*,
https://doi.org/10.1007/978-981-19-9715-0_3

to make policies with scientific methods; enterprises can maximize profits; scholars can find scientific laws to optimize social and economic development.

In March 2014, it was the first time Big Data was recorded in the Government Work Report and mentioned on several occasions by Premier LI Keqiang as a "buzzword". Big data plays a strategic role in national governance. Some local governments of cities and provinces such as Shanghai, Guangdong, Guizhou, and Shaanxi were keenly aware of the upcoming opportunities of big data, putting forward plans of developing big data and building associated local industries.

In March 2015, Premier LI Keqiang proposed "Internet+" and the strategic importance of big data application at the third session of the 12th National People's Congress and the 10th National Committee for the Chinese People's Political Consultative Conference, receiving reactions across the country. People began to realize the true potential of the Internet and big data, which can be used at all levels of national governance, including national security, government statistics, economic forecasting, public opinion monitoring, as well as a financial investment and social research, and more often in government management decisions, such as prior decision support, midterm supervision and post-evaluation.

In October 2015, the construction of big data has reached to a new stage when it was first written into the Party's plenum resolution at the 5th Session of the 18th CPC Central Committee Plenary. Big data has officially become a national strategy as proposed in the communique to implement the "National Big Data Strategy". Since then, regional and local departments have launched policies and measures related to the development of big data, through which big data has been rapidly promoted.

At present, big data in China is entering a development stage. Big data will provide broader and stronger support for social and economic development, and is expected to drive the transformation of the IT service industry with a market scale of trillion yuan during the "13th Five-Year Plan". With the introduction and implementation of national policies and promotion of local governments' sharing and opening of government information resources, the application of big data in government affairs will deepen gradually, becoming an important method of macroeconomic regulation and control, market supervision, and the effectiveness of public services, which strongly support the improvement of the effectiveness of government administrative services and the optimization of social governance. Big data in government affairs will further enhance the effectiveness of government services and the social governance level.

2 Development Level Index of Big Data in China

2.1 The General Development Trend Is in a Good Position

2.1.1 Steady and Progressive Development of Big Data Industry

As shown in Fig. 1, in the first half of 2017, the development level index of big data in China was 65.11, which is steady and progressive.

In the first half of 2017, the fundamental index of big data in China was 70.58, which was an increase compared to 68.20 in the previous period, showing the fundamental support for the development of big data industry has become increasing solid. In the first half of 2017, China's GDP was 38.15 trillion yuan, indicating a 6.9% year-on-year at comparable prices. Faced with the international economic situation of sluggish growth and the pressure brought by the transformation and upgrading of domestic industrial structure, the overall stable and steady trend of national economic operation shows that the economic foundation is even more sufficient to support the development of big data industry in China. Besides, in recent years, the amount of Chinese high-quality talents has steadily increased. In 2016, the total scale of all kinds of higher education in the country reached 36.99 million, and the gross enrollment rate of higher education reached 42.7%, among which, there were 342,000 Ph.D. students and 1,639,000 postgraduate students, which laid a good intellectual foundation for the development of big data industry.

In the first half of 2017, the stimulus index of big data in China was 59.64, which was increased compared to 57.51 in the previous period, showing the factors stimulating the development of big data industry have been continuously optimized. Policy guidance in the development of big data industry has been enhanced, creating a favorable policy environment. Policy support is also enhanced across provinces, promoting

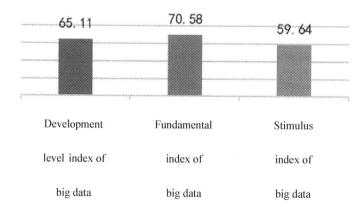

Fig. 1 Development level index of big data in China

the leap development of big data industry from storage, computing, talents, applications, and other related areas. In addition, those local governments are also actively using various forms and channels to promote the development and practical application of big data, creating a good public opinion environment for the development of the industry.

2.1.2 The Elements of Big Data Industry Development Are Still Uneven

As shown in Fig. 2, in terms of the composition of the index, the fundamental and stimulus index takes up 54.2 and 45.8% respectively of the development level index of big data. Although the composition is still uneven, the proportion of the stimulus compared has increased slightly compared to the previous period, indicating that the targeted fields and measures for promoting the development of big data industry are enhanced across provinces. As they may have optimized the layout of associated industries, cultivated talents in related disciplines, refined policy support in fields, guided the media to promote reasonably, etc.

2.2 Beijing and Guangdong Are Leading in the Development Level of Big Data

As shown in Fig. 3, in the provincial development level index of big data of the first half of 2017, the indices of Beijing, Guangdong, Shanghai, Zhejiang, Jiangsu, Shandong, Sichuan, Guizhou are all above 70, which are 83.17, 78.69, 78.24, 76.63, 72.77, 71.21, 70.76, 70.31 respectively; indices of Tianjin, Hubei, Fujian, Hebei, Henan are 69.95, 69.24, 66.92, 66.48, 65.88 respectively that lower than 70 but higher than the national level at 65.11; indices of Liaoning, Chongqing, Shaanxi,

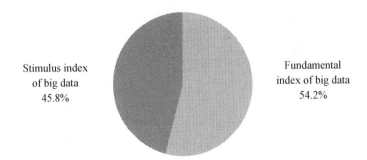

Fig. 2 Composition of development level index of big data in China

Sketch of the Development of Big Data in China

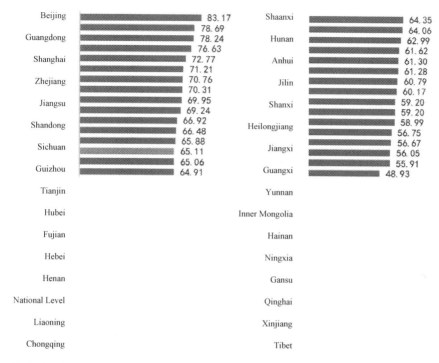

Fig. 3 Provincial development level index of big data

Hunan, Anhui, Jilin, Shanxi, Heilongjiang, Jiangxi, Guangxi are higher than 60 but lower than the national level, which are 65.06, 64.91, 64.35, 64.06, 62.99, 61.62, 61.30, 61.28, 60.79, 60.17 respectively; of the rest are lower than 60, and of Tibet is lower than 50. Overall, among the 31 provinces, there are 13 show in index above the national level, consisting of 41.9%, which is lower than the previous period and indicates an increasing trend of the development gap between provinces.

2.2.1 There Are 29 Provinces That Show Fundamental Support Better than the Stimulus Factor

As the composition shown in Fig. 4, regions with a high development level of big data industry index are relatively balanced at fundamental index and stimulus index; while regions with a low index shows a larger difference between the two indices. In terms of proportions, among the 31 provinces, the fundamental index of Sichuan and Guizhou are lower than the stimulus index there. Sichuan Province has the smallest gap, indicating that it is more pragmatic in the development of big data and

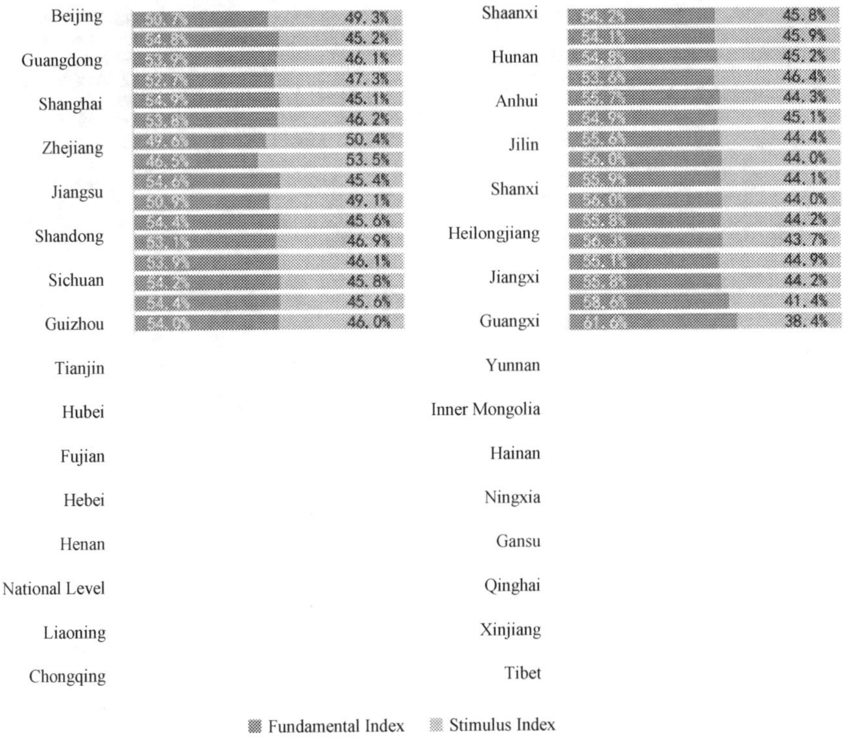

Fig. 4 Composition of provincial development level index of big data

Legend: ▓ Fundamental Index ▒ Stimulus Index

Province	Fundamental Index	Stimulus Index
Beijing	50.7%	49.3%
	54.8%	45.2%
Guangdong	53.9%	46.1%
	52.7%	47.3%
Shanghai	54.9%	45.1%
	53.8%	46.2%
Zhejiang	49.6%	50.4%
	46.5%	53.5%
Jiangsu	54.6%	45.4%
	50.9%	49.1%
Shandong	54.4%	45.6%
	53.1%	46.9%
Sichuan	53.9%	46.1%
	54.2%	45.8%
Guizhou	54.4%	45.6%
	54.0%	46.0%
Shaanxi	54.2%	45.8%
	54.1%	45.9%
Hunan	54.8%	45.2%
	53.6%	46.4%
Anhui	55.7%	44.3%
	54.9%	45.1%
Jilin	55.6%	44.4%
	56.0%	44.0%
Shanxi	55.9%	44.1%
	56.0%	44.0%
Heilongjiang	55.8%	44.2%
	56.4%	43.7%
Jiangxi	55.1%	44.9%
	55.8%	44.2%
Guangxi	58.6%	41.4%
	61.6%	38.4%

Other provinces (left column): Tianjin, Hubei, Fujian, Hebei, Henan, National Level, Liaoning, Chongqing.

Other provinces (right column): Yunnan, Inner Mongolia, Hainan, Ningxia, Gansu, Qinghai, Xinjiang, Tibet.

that the factors promoting the development of big data are better integrated with its own economic and social development. Among the regions with high indexes, such as Beijing, Guangdong, Shanghai, Zhejiang, and Jiangsu, Beijing shows the best balance with little difference between its fundamental and stimulus index; Hubei has a similar structure as Beijing; followed by Zhejiang; and then Shanghai, Guangdong, and Jiangsu. Tibet shows the biggest difference where the fundamental index reaches 61.6%.

2.2.2 The Development Level Index of Big Data Is Relatively Higher in Economically Developed Regions

The provinces with a high level of the development level index are mainly located in the Yangtze River Delta, the Pearl River Delta, and the Bohai Sea region, with Sichuan and Guizhou in the southwest also standing out.

2.2.3 Guangdong, Shanghai, Beijing, and Zhejiang Have Laid a Solid Foundation for Industrial Development

In the first half of 2017 as shown in Fig. 5, the fundamental indices of Guangdong, Shanghai, Beijing, and Zhejiang all exceeded 80, at 86.24, 84.39, 84.29, 80.75 respectively; the fundamental indices of Jiangsu, Shandong, Tianjin, Fujian, Henan, Liaoning, Hebei, Hubei, Chongqing, and Sichuan are 79.96, 76.66, 76.39, 72.81, 70.98, 70.77, 70.62, 70.55, 70.16, 70.16 respectively, which are higher than 70 but lower than 80; there are in total 20 provinces, consist of 64.5% of all 31 provinces have the index below the national level, among which Hubei, Sichuan, and Chongqing are not far behind the national level.

As the composition of the fundamental index shown in Fig. 6, Shandong shows a more even weighting of all dimensions and the best balance; Shanghai, with its strengths in ports, air logistics, and foreign trade, has the highest efficiency level in the country with a weighting of 26.2%; Jiangsu has a more prominent scale level with a weighting of 29.0%; Beijing and Zhejiang have a similar composition of the

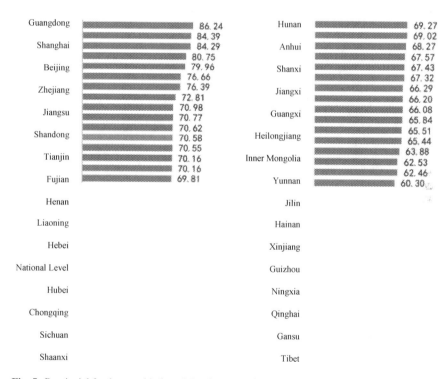

Fig. 5 Provincial fundamental index of development of big data

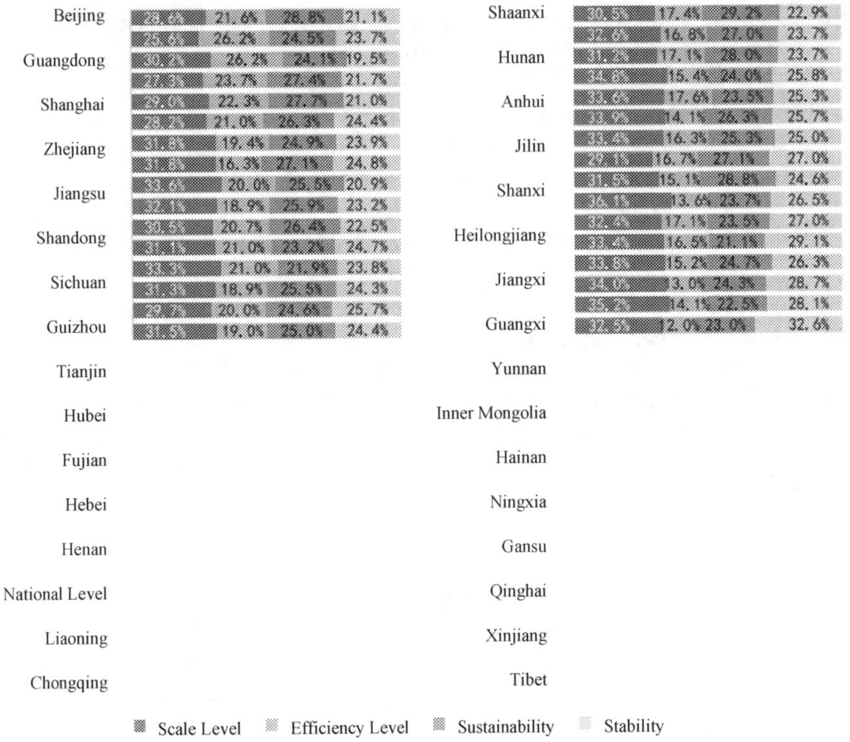

Beijing 21.6% 28.8% 21.1%
26.2% 24.5% 23.7%
Guangdong 26.2% 24.1% 19.5%
23.7% 27.4% 21.7%
Shanghai 22.3% 27.7% 21.0%
21.0% 26.3% 24.4%
Zhejiang 19.4% 24.9% 23.9%
16.3% 27.1% 24.8%
Jiangsu 20.0% 25.5% 20.9%
18.9% 25.9% 23.2%
Shandong 20.7% 26.4% 22.5%
21.0% 23.2% 24.7%
Sichuan 21.0% 21.9% 23.8%
18.9% 25.5% 24.3%
Guizhou 20.0% 24.6% 25.7%
19.0% 25.0% 24.4%
Tianjin
Hubei
Fujian
Hebei
Henan
National Level
Liaoning
Chongqing

Shaanxi 17.4% 29.2% 22.9%
16.8% 27.0% 23.7%
Hunan 17.1% 28.0% 23.7%
15.4% 24.0% 25.8%
Anhui 17.6% 23.5% 25.3%
14.1% 26.3% 25.7%
Jilin 16.3% 25.3% 25.0%
16.7% 27.1% 27.0%
Shanxi 15.1% 28.8% 24.6%
13.6% 23.7% 26.5%
Heilongjiang 17.1% 23.5% 27.0%
16.5% 21.1% 29.1%
Jiangxi 15.2% 24.7% 26.3%
13.0% 24.3% 28.7%
Guangxi 14.1% 22.5% 28.1%
12.0% 23.0% 32.6%
Yunnan
Inner Mongolia
Hainan
Ningxia
Gansu
Qinghai
Xinjiang
Tibet

Scale Level   Efficiency Level   Sustainability   Stability

Fig. 6 Composition of provincial fundamental index of development of big data

fundamental index with higher weightings for scale level and sustainability, and equal weightings for efficiency level and stability; Guangdong has equal weightings for scale level and efficiency level, and equal weightings for sustainability and stability.

2.2.4 Beijing, Guizhou, and Zhejiang Vigorously Promote the Industrial Development

As shown in Fig. 7, the stimulus indices of Beijing, Guizhou, Zhejiang, Shanghai, Sichuan, and Guangdong are all above 70, at 82.05, 75.17, 72.51, 72.10, 71.36, 71.14 respectively; indices of Hubei, Shandong, Jiangsu, Tianjin, Hebei, Fujian, and Henan are 67.93, 65.77, 65.57, 63.50, 62.34, 61.03, 60.7 respectively, which are higher than 60 but lower than70; that of Chongqing is lower than 60 but higher than the national level at 59.64; index of Liaoning, Shaanxi, Hunan, Jilin, Anhui, Heilongjiang, Shanxi, Jiangxi, Guangxi, Yunnan, Hainan, Inner Mongolia, and Gansu

Sketch of the Development of Big Data in China

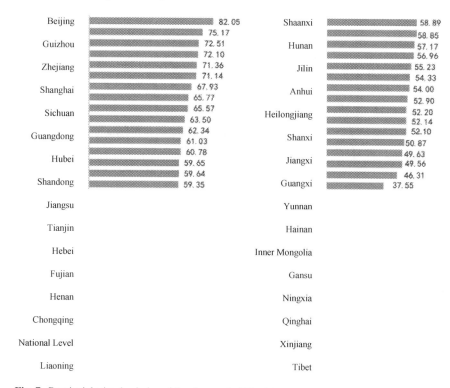

Fig. 7 Provincial stimulus index of development of big data

are lower than the national level but higher than 50; indices of the rest provinces are lower than 50, of which, index of Tibet is lower than 40, at 37.55. Overall, there are 14 provinces, accounting for 45.2% of the total, have the stimulus index higher than the national level; the index of Beijing is more than two times that of Tibet, indicating that there are large differences among provinces in terms of hardware (i.e., natural endowments such as the gathering of related industries and talent reserves) and software (i.e. related policy measures and public opinion propaganda) for promoting the development of big data industry.

As the composition of the stimulus index is shown in Fig. 8, there are different characteristics among provinces. Comparing all factors, Beijing has the highest proportion of the industrial environment at 29.5%, while the talent reserve is similar at 29.4%; Jiangsu has a more prominent proportion of the talent reserve at 34.0%, followed by the policy measures and the industrial environment at 29.3 and 28.7% respectively; the policy measures have the highest weighting in Qinghai at 50.9%, while Guizhou has the highest weighting in the country for the public opinion propaganda at 33.4%, followed by the policy measures dimension and the talent reserve

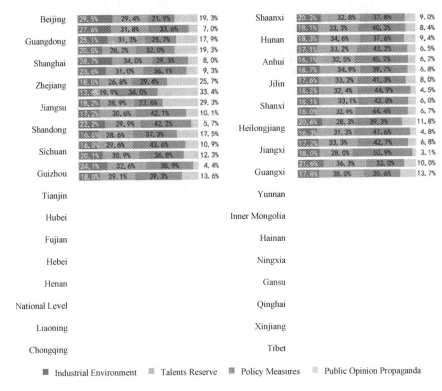

Fig. 8 Composition of provincial stimulus index of development of big data

at 34.0 and 19.9% respectively. Overall, the central tendency becomes even more obvious along with the decreasing of the stimulus index, with occasions that the individual weighting approaches or exceeds 50%.

2.3 The Development Level Is Higher in Shenzhen and Guangzhou

In the first half of 2017 as shown in Fig 9, among major cities, the development level indices of big data of Shenzhen and Guangzhou are higher, at 81.28 and 79.06 respectively; the indices of Wuhan, Chengdu, Nanjing, and Hangzhou are higher than 70, at 75.44, 74.81, 72.47, 72.00 respectively; the indices of Xi'an, Qingdao, Jinan, Changsha, Zhengzhou, Xiamen, Changchun, Ningbo, Suzhou, Guiyang, and Fuzhou are higher than 60 but lower than 70, at 68.29, 67.66, 65.78, 65.15, 63.99, 63.54, 63.29, 63.25, 62.83, 62.24, 61.15 respectively.

Sketch of the Development of Big Data in China

City	Value	City	Value	City	Value
Shenzhen	81.28 / 79.06	Foshan	57.15 / 57.13	Zunyi	48.98 / 48.50
Guangzhou	75.44 / 74.81	Harbin	56.22 / 55.52	Guilin	48.23 / 48.11
Wuhan	72.47 / 72.00	Nantong	54.98 / 54.48	Jiujiang	47.98 / 47.70
Chengdu	68.29 / 67.66	Shijiazhuang	53.91 / 53.55	Zhangjiajie	47.32 / 47.16
Nanjing	65.78 / 65.15	Yantai	53.47 / 53.45	Yueyang	46.36 / 45.57
Hangzhou	63.99 / 63.54	Nanchang	53.42 / 53.34	Jining	45.46 / 45.27
Xi'an	63.29 / 63.25	Yinchuan	53.02 / 52.54	Xining	45.07 / 44.97
Qingdao	62.83 / 62.24	Nanning	52.46 / 52.09	Baoding	44.65 / 44.45
Jinan	61.15 / 59.85	Jilin	52.07 / 51.98	Xiangyang	44.37 / 43.51
Changsha	59.47 / 59.32	Baotou	51.97 / 51.78	Yichang	43.35 / 43.07
Zhengzhou	59.30 / 59.17	Lanzhou	51.69 / 51.56	Ganzhou	42.54 / 42.32
Xiamen	58.65 / 58.52	Jinhua	50.06 / 49.76	Zhanjiang	41.95 / 41.50
Changchun	58.03 / 57.30 / 57.29	Tangshan	49.41 / 49.30 / 49.10	Anqing	41.15 / 41.14 / 39.51
Ningbo		Luoyang		Huai'an	
Suzhou		Sanya		Shaoguan	
Guiyang		Yangzhou		Changde	
Fuzhou		Huizhou		Jiaozuo	
Quanzhou		Zibo		Bengbu	
Kunming		Hohhot		Lhasa	
Taiyuan		Zhenjiang		Dali	
Dalian		Weifang		Pingdingshan	
Hefei		Haikou		Jinzhou	
Shenyang		Xuzhou		Nanchong	
Wuxi		Beihai		Cangzhou	
Urumqi		Qinhuangdao		Dandong	
Dongguan		Luzhou		Enshi	
Wenzhou		Yancheng		Mudanjiang	

Fig. 9 Development level index of big data in major cities

2.3.1 Most Cities Have More Stimulus Factors than Fundamental Support

As the composition of the development level index of big data of major cities is shown in Fig. 10, there are 42 cities show higher weighting of the index of stimulus

City	Fundamental Support	Stimulus Factor
Shenzhen	49.1%	50.9%
	45.1%	54.9%
Guangzhou	44.8%	55.2%
	41.9%	58.1%
Wuhan	45.7%	54.3%
	48.4%	51.6%
Chengdu	48.7%	51.3%
	46.0%	54.0%
Nanjing	46.1%	53.9%
	47.4%	52.6%
Hangzhou	45.6%	54.4%
	47.8%	52.2%
Xi'an	55.0%	45.0%
	48.3%	51.7%
Qingdao	53.9%	46.1%
	47.5%	52.5%
Jinan	46.5%	53.5%
	55.2%	44.8%
Changsha	49.4%	50.6%
	48.0%	52.0%
Zhengzhou	48.7%	51.3%
	46.9%	53.1%
Xiamen	43.6%	56.4%
	51.0%	49.0%
Changchun	53.4%	46.6%
	49.5%	50.5%
Ningbo		
Suzhou		
Guiyang		
Fuzhou		
Quanzhou		
Kunming		
Taiyuan		
Dalian		
Hefei		
Shenyang		
Wuxi		
Urumqi		
Dongguan		
Wenzhou		
Foshan	53.8%	46.2%
	45.2%	54.8%
Harbin	54.9%	45.1%
	46.5%	53.5%
Nantong	51.9%	48.1%
	48.4%	51.6%
Shijiazhuang	52.4%	47.6%
	48.5%	51.5%
Yantai	43.5%	56.5%
	54.5%	45.5%
Nanchang	47.3%	52.7%
	57.8%	42.2%
Yinchuan	49.2%	50.8%
	49.6%	50.4%
Nanning	49.4%	50.6%
	51.8%	48.2%
Jilin	51.2%	48.8%
	48.3%	51.7%
Baotou	52.2%	47.8%
	53.0%	47.0%
Lanzhou	55.9%	44.1%
	46.3%	53.7%
Jinhua	50.8%	49.2%
	51.8%	48.2%
Tangshan	53.6%	46.4%
	51.3%	48.7%
	50.8%	49.2%
Luoyang		
Sanya		
Yangzhou		
Huizhou		
Zibo		
Hohhot		
Zhenjiang		
Weifang		
Haikou		
Xuzhou		
Beihai		
Qinhuangdao		
Luzhou		
Yancheng		
Zunyi	49.6%	50.4%
	47.4%	52.6%
Guilin	50.3%	49.7%
	45.1%	54.9%
Jiujiang	53.8%	46.2%
	53.0%	47.0%
Zhangjiajie	52.3%	47.7%
	48.2%	51.8%
Yueyang	51.3%	48.7%
	53.5%	46.5%
Jining	50.7%	49.3%
	49.0%	51.0%
Xining	50.6%	49.4%
	51.3%	48.7%
Baoding	50.1%	49.9%
	50.2%	49.8%
Xiangyang	50.8%	49.2%
	52.3%	47.7%
	47.4%	52.6%
Yichang	49.9%	50.1%
	47.4%	52.6%
Ganzhou	49.8%	50.2%
	50.3%	49.7%
Zhanjiang	45.1%	54.9%
	52.2%	47.8%
Anqing	51.0%	49.0%
	52.6%	47.4%
Huai'an		
Shaoguan		
Changde		
Jiaozuo		
Bengbu		
Lhasa		
Dali		
Pingdingshan		
Jinzhou		
Nanchong		
Cangzhou		
Dandong		
Enshi		
Mudanjiang		

▨ Fundamental Support   ▨ Stimulus Factor

Fig. 10 Composition development level index of big data in major cities factors than that of the fundamental support, accounting for 51.9% of the 81 cities covered in this study. It all shows that the proportion of stimulus factor is higher than that of fundamental support in the top 10 cities in terms of the development level of big data, indicating that these cities have taken many measures to promote the development with the support of their own benign foundation.

2.3.2 Shenzhen Has Laid a Good Foundation for the Development of Big Data

As shown in Fig. 11, Shenzhen is outstanding with a fundamental index of 79.80, while that of Guangzhou is 71.36, both higher than 70. This indicates that the own foundation of both two cities can provide a good support platform for the development of big data. On the other hand, all four dimensions of fundamental support of Shenzhen are relatively high and balanced.

2.3.3 Chengdu Vigorously Promotes Development of Big Data

As shown in Fig. 12, in terms of stimulus index, Chengdu, Guangzhou, Wuhan, and Shenzhen are higher than other cities, at 86.87, 86.75, 83.30, and 82.77 respectively. By further analyzing the composition of the stimulus index for the development of big data industry in cities, it is found that Chengdu is better than Shenzhen in terms of the industrial environment; Guangzhou is better than Chengdu in terms of talent reserves, policy measures and public opinion propaganda; Wuhan has certain advantages in terms of talent reserve; Shenzhen is more prominent in terms of public opinion propaganda; while Nanjing, Hangzhou, Qingdao, Jinan, and Xi'an have a stimulus index between 70 and 80. Nanjing and Hangzhou are similar in terms of the composition of the stimulus index, as both cities excel in the dimension of public opinion propaganda and are insufficient in the dimension of the industrial environment related to big data.

Research Notes

With thinking, analysis, and application of big data, this study adopts big data technology to obtain, process, analyze, and mine relevant internet data; combines local statistical data and government data from different areas; and uses the deviation standardization method to process the data. The aim is to quantify and analyze the development level of big data industry in 31 provinces through the development level index of big data industry, and study the current situation of its development in each province. Through this, we discovered the strengths and weaknesses of each area, so as to help the local government promote and guide related work.

City	Value	City	Value	City	Value
Shenzhen	79.80	Yantai	57.09	Xining	49.49
	71.36		56.94		48.73
Guangzhou	69.68	Taiyuan	56.90	Yichang	48.61
	69.65		56.66		48.51
Changchun	67.68	Fuzhou	56.47	Zunyi	47.74
	67.59		55.48		47.54
Hangzhou	66.56	Wenzhou	54.90	Jiujiang	46.57
	66.17		54.26		46.15
Suzhou	66.03	Yinchuan	53.95	Haikou	46.13
	62.76		53.30		46.02
Wuhan	62.23	Hefei	52.94	Xiangyang	45.66
	62.03		52.74		45.50
Xi'an	61.81	Zhenjiang	52.20	Jilin	45.48
	61.76		52.12		45.08
Nanjing	61.70	Huhehaote	51.90	Huai'an	44.77
	61.54		51.83		44.60
Quanzhou	61.12	Yangzhou	51.66	Ganzhou	44.37
	60.77		51.63		43.36
Chengdu	60.67	Huizhou	51.63	Guilin	42.99
	59.91		51.55		42.98
Qingdao	59.71	Qinhuangdao	51.14	Anqing	42.17
	59.14		50.82		42.14
Wulumuqi	58.73	Nanchang	50.59	Bengbu	41.96
	58.32		50.55		41.56
Changsha	58.31	Tangshan	50.52	Baoding	41.13
	57.80		50.20		40.32
	57.77		49.87		37.40
Nantong		Luoyang		Jiaozuo	
Jinhua		Nanning		Shaoguan	
Foshan		Sanya		Changde	
Ningbo		Shijiazhuang		Zhanjiang	
Xiamen		Yueyang		Zhangjiajie	
Jinan		Harbin		Dandong	
Dongguan		Beihai		Dali	
Wuxi		Shenyang		Nanchong	
Guiyang		Xuzhou		Jinzhou	
Kunming		Luzhou		Enshi	
Zhengzhou		Lanzhou		Mudanjiang	
Baotou		Jining		Lhasa	
Dalian		Zibo		Pingdingshan	
Weifang		Yancheng		Cangzhou	

Fig. 11 Fundamental index of development of big data in major cities

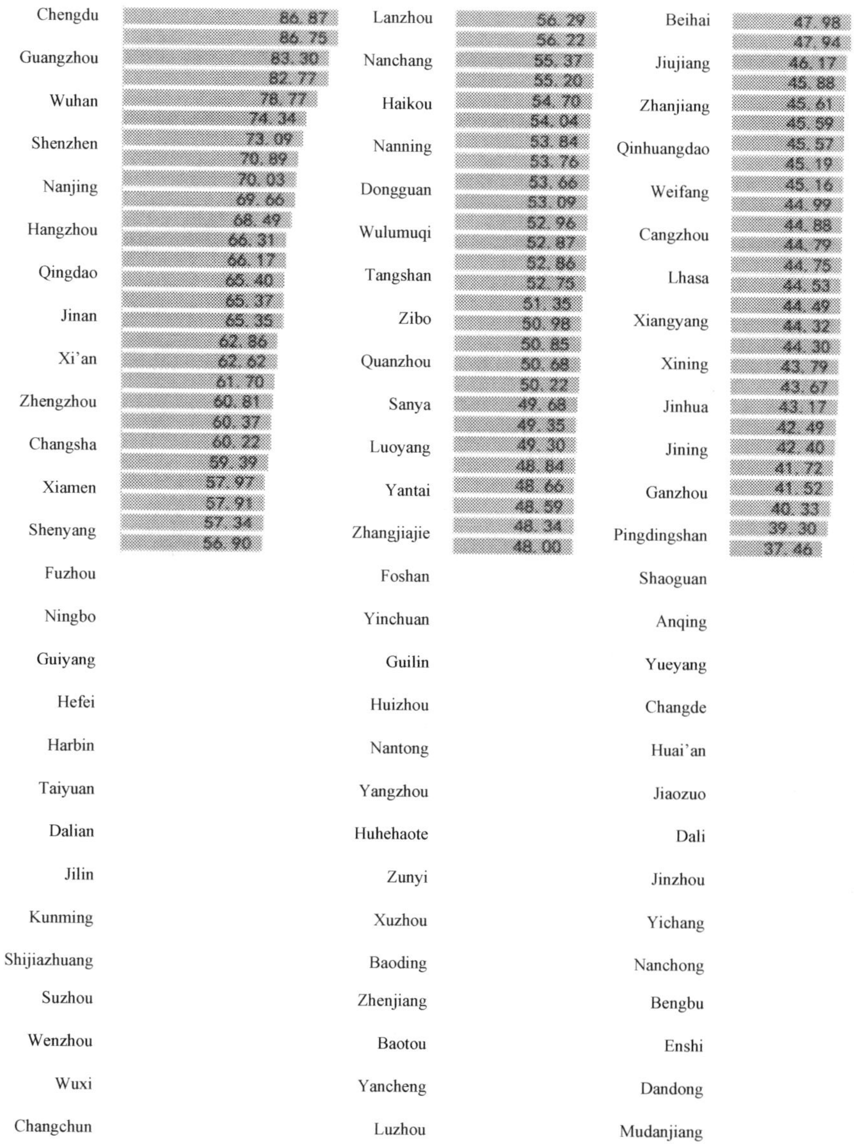

Chengdu 86.87 / 86.75
Guangzhou 83.30 / 82.77
Wuhan 78.77 / 74.34
Shenzhen 73.09 / 70.89
Nanjing 70.03 / 69.66
Hangzhou 68.49 / 66.31
Qingdao 66.17 / 65.40
Jinan 65.37 / 65.35
Xi'an 62.86 / 62.62
Zhengzhou 61.70 / 60.81
Changsha 60.37 / 60.22
Xiamen 59.39 / 57.97
Shenyang 57.34 / 56.90
Fuzhou
Ningbo
Guiyang
Hefei
Harbin
Taiyuan
Dalian
Jilin
Kunming
Shijiazhuang
Suzhou
Wenzhou
Wuxi
Changchun

Lanzhou 56.29 / 56.22
Nanchang 55.37 / 55.20
Haikou 54.70 / 54.04
Nanning 53.84 / 53.76
Dongguan 53.66 / 53.09
Wulumuqi 52.96 / 52.87
Tangshan 52.86 / 52.75
Zibo 51.35 / 50.98
Quanzhou 50.85 / 50.68
Sanya 50.22 / 49.68
Luoyang 49.35 / 49.30
Yantai 48.84 / 48.66
Zhangjiajie 48.59 / 48.34 / 48.00
Foshan
Yinchuan
Guilin
Huizhou
Nantong
Yangzhou
Huhehaote
Zunyi
Xuzhou
Baoding
Zhenjiang
Baotou
Yancheng
Luzhou

Beihai 47.98 / 47.94
Jiujiang 46.17 / 45.88
Zhanjiang 45.61 / 45.59
Qinhuangdao 45.57 / 45.19
Weifang 45.16 / 44.99
Cangzhou 44.88 / 44.79
Lhasa 44.75 / 44.53
Xiangyang 44.49 / 44.32
Xining 44.30 / 43.79
Jinhua 43.67 / 43.17
Jining 42.49 / 42.40
Ganzhou 41.72 / 41.52
Pingdingshan 40.33 / 39.30 / 37.46
Shaoguan
Anqing
Yueyang
Changde
Huai'an
Jiaozuo
Dali
Jinzhou
Yichang
Nanchong
Bengbu
Enshi
Dandong
Mudanjiang

Fig. 12 Stimulus index of development of big data in major cities

A total of 31 provinces and 81 major cities across the country were selected for this study.

The data period for this study is from 1 January 2017 to 30 June 2017.

Components of the Development Level Index System of Big Data Industry

Big Data Index. A quantitative analysis of fundamental support and operating conditions of the development of big data in provinces and major large and medium-sized cities across China, reflecting the specific applications of current status and level of the development of big data industry in each province and major large and medium-sized cities. If to compare the development system of big data industry to a running machine, its index is the dashboard reflecting the running condition of the machine.

The Fundamental Support for the Development of Big Data Industry. The fundamental part of the development reflects the foundation or starting point level of big data industry. The higher this fundamental platform is, the higher the foundation level of development of big data industry is, and the higher the starting point is. The research group mainly describes and portrays the fundamental platform system of development from the perspective of scale level, efficiency level, sustainability, and stability.

Stimulus Factors for the Development of Big Data Industry. The main part of the development reflects the main factors and measures that enhance and promote the development of big data industry. The better and more comprehensive the factors and measures are, the smoother and healthier the development will be. The research team mainly describes and measures the stimulus factors that promote the development from the elements of the industrial environment, talent reserve, policy measures, and public opinion propaganda.

Reported by: Research Laboratory of Big Data of CSISC

3 Jointly Established Research Institute to Promote the Construction of "Double First-Class" Universities

In the context of economic globalization and China's comprehensively deepening reform, talent development has to be based on facts, evidence, and data. However, we find that talent development today suffers from a lack of data, not only in terms of insufficient data volume but more importantly in terms of a single dimension of data that is too structured, while there is basically nothing in the real big data on talent and growth, which greatly limits talent selection, specialized education, career development decisions as well as reform and innovation.

With the need for national strategy and proactively responding to the important deployment of the era of big data, the future of China's university education should develop in the way to master and go beyond big data. On 18 August 2015, the

15th meeting of the Central Leading Group for Comprehensively Deepening Overall Reform considered and adopted the "Overall Plan for Coordinating the Construction of World-Class Universities and First-Class Disciplines" and decided to coordinate the construction of world-class universities and first-class disciplines.

In September 2017, the Ministry of Education, the Ministry of Finance, and the National Development and Reform Commission jointly issued the Notice on the Announcement of the List of Universities and Construction Disciplines for the Construction of World-Class Universities and First-Class Disciplines, which is seen as an official confirmation and announcement.

The world-class universities and first-class disciplines are referred to as "double first-class". The construction of world-class universities and first-class disciplines is a major strategic decision made by the Central Committee of the Party and the State Council, which is conducive to enhancing the overall strength and international competitiveness of China's higher education and providing strong support for the realization of the Two Centenary Goals and the Chinese Dream of the great rejuvenation of the Chinese nation. The construction of "double first-class" is an open system to all universities, even higher education institutions. The "double first-class" policy will dominate the development of higher education in China for a long time to come. At present, it is difficult for local institutions to compete with the "985" and "211" universities, but it does not mean that they do not have opportunities in all disciplines; in the future, as long as local institutions seize the historical opportunities, formulate scientific development strategies, and choose the right path, it is entirely possible for them to make a local institution to become outstanding, building the first-class disciplines and becoming a world-class university. The model of co-establishing research institutes based on big data is a new model for the construction of "double first-class" universities to explore the cultivation of core competitiveness and the solution of talent problems in provincial universities, which is market-led, data-led, and data-open sharing, promoting the deep integration of industry, university and research. This model is conducive to cultivating a multi-level and multi-type talent team for big data, as well as fostering the creation of leading big data enterprises, and is also a practical exploration of a data-driven innovation system for the construction of "double first-class".

3.1 Chinese Academy of Education Big Data

The Chinese Academy of Education Big Data is an institute jointly built in September 2015 by the China Statistical Information Service Centre (CSISC) and the Qufu Normal University in the hometown of Confucius. Its research focuses on national public opinion tracking of major education policies, basic education quality monitoring, educational targeted poverty alleviation based on big data, university governance, and performance evaluation, national identity and education strategies of fine traditional culture, physical health monitoring of students in primary and secondary schools and universities, analysis of students' cognitive tendencies and professional development trends, and comparative research on mathematics, reading, and science literacy and on education teaching strategies of primary and secondary school students in China and abroad. In December 2015, the institute led more than ten universities and educational research institutions to launch the "Development and Promotion Plan of Chinese Education on Big Data", first proposing a "road map" for promoting the development of big data in China's education; built the first "Experience Center of Omni Media and Big Data Education" in China. The China Education Daily published an in-depth interview with QI Wanxue, President of the Institute, and me on the topic of the "Targeted Treatment" for Education Reform in the Era of Big Data, and made a full-page report.

In September 2015, the Chinese Academy of Education Big Data was jointly built by the China Statistical Information Service Centre and the Qufu Normal University in the hometown of Confucius

3.2 China (Xi'an) Institute for Silk Road Research | Big Data Research Center of "Belt and Road"

On 17 January 2017, the signing and inauguration ceremony of China (Xi'an) Institute for Silk Road Research and Big Data Research Center of "Belt and Road" was held in Beijing. HU Jian, President of Xi'an University of Finance and Economics, DING Jutao, Vice President of Xi'an University of Finance and Economics, YAN Jianhui, Director of CSISC, WANG Haifeng, Deputy Director of CSISC, CHEN

Wei, Vice Chairman of Home Big Data, and LIANG Xiaojie, Director of Chinaso attended the ceremony that I presided over. The research center is jointly built and run by the CSISC, Xi'an University of Finance and Economics [China (Xi'an) Institute for Silk Road Research], and the Home Big Data company, which is another pragmatic project as a milestone for the implementation of the "Belt and Road" Initiative after the China (Xi'an) Institute for Silk Road Research jointly built by National Bureau of Statistics and Shaanxi Provincial People's Government. The research area is focused on data integration, data mining, and technical algorithm development and application for countries and provinces along the "Belt and Road", with a view to absorbing and integrating data from the whole region, so as to enrich "The National Big Data Pool Project". The research center will also share research results based on big data methods and technologies, and provide a big data think tank for theoretical innovation and practical application of the "Belt and Road". HU Jian said that China (Xi'an) Institute for Silk Road Research | Big Data Research Center of "Belt and Road" would be committed to creating its own unique think-tank products on "Belt and Road", which would provide a reference for national and local development; integrating and mining domestic and oversea data to upgrade management information systems into decision support systems so as to improve decision support; as well as striving to achieve the goal of transforming planning into real project and countermeasures into policies and to develop a path with its own characteristics to avoid homogenization.

On 17 January 2017, the signing and inauguration ceremony of China (Xi'an) Silk Road Research Institute and Big Data Research Center of "Belt and Road" was held in Beijing

YAN Jianhui said that the signing of Big Data Research Center of "Belt and Road" is a concrete action of the center to implement the national "Belt and Road" initiative, and it should fully utilize the advanced management concept, superb advantages on scientific research, and strong expert team of Xi'an University of Finance and Economics to do a good job in supporting the research and application of big data and make new contributions to the "Belt and Road" initiative and the development strategy of big data.

CHEN Wei believes that it is a sacred mission for Home Big Data to undertake the specific work of Big Data Research Center of "Belt and Road". In the critical transition period of the global economy which includes the Chinese economy, people's demand for timely and accurate information is getting stronger and stronger, while the service industry proportion is becoming bigger and bigger. The popularity of the Internet is allowing information to be transmitted to users in the manner of flat design, which will be a huge change and development trend in human economic life. In such a period of transition, new ways of thinking and new technologies are needed to promote people's interaction with data and to provide high-quality data information services for governments and enterprises through data mining. The establishment of Big Data Research Center of "Belt and Road" has both historical and practical significance at present, as well as profound significance in the long term.

4 Promoting the Comprehensive Integration of Big Data and Economic Society through Base Cultivation

Big data is becoming a strategic focus and a high ground in the new round of technological and industrial competition. Both domestic and international government departments and industries are considering big data as a strategic emerging industry and a driving force for the upgrading and transformation of traditional industries, and have accelerated the planning deployment and promoted the development of big data applications. With the gradual improvement of the local infrastructure of big data, creating value through its applications in government and industries is becoming a top priority for the development of big data across the country. After more than a year of preparation, the first Big Data Research Service Base jointly built by CSISC and Xiamen City was inaugurated in Xiamen on 28 June 2014, while Big Data Research Laboratory was also settled in the Cyphy Technology Co., Ltd (Xiamen). Big Data Research Service Base will build a platform for the development of big data industry, integrate resources from all fields of industry, academia, and research, and accelerate the development of big data industry in Xiamen and even China. In the speech made by XU Wendong, Deputy Secretary General of Xiamen Municipal People's Government, on behalf of the municipal government, he stressed that

Xiamen will vigorously develop big data industry and other associated industries, strengthen cooperation with CSISC, support the construction of Big Data Research Service Base and Big Data Research Laboratory, and strengthen the software and information service industry in Xiamen. Meanwhile, Xiamen will also take the lead in opening up government data resources, and planning and building the Xiamen Data Portal, so that to encourage social capital to invest in the area, giving rise to more and wider applications of big data. KONG Shuguang, Director of the Xiamen Information Bureau at the time (current Director of the Xiamen Bureau of Statistics), said that Xiamen has already established a relatively complete information resource system. Hope in the future cooperation with the CSISC, they can combine national-level big data research results with local economic development needs to create a new advantage in the development of Xiamen's software industry, thereby attracting more enterprises and organizations to participate in the development of Xiamen's big data industry, thus to jointly promote the development of Xiamen's big data industry clusters.[1]

On 28 June 2014, the first Big Data Research Service Base jointly built by CSISC and Xiamen City was inaugurated in Xiamen

---

[1] WANG Yali. A New Industrial Model to Help the Construction of a Famous Software City [N]. China Quality Daily, 2014:07–07.

The service base model of big data research is the most typical development model for applications of big data. The base is a pioneering area for research and application of big data with regional characteristics, as well as a cluster area for big data entrepreneurship and innovation. Through strategic measures, such as integrating the advantages of regional industrial resources through national data, cooperative research and development of technologies, regional industrial convergence and affiliated support, local talent training, and investment incubation, the base can realize the upgrading and transformation of local government and industrial economy, especially of the real economy with a new big data model, promoting the development of regional characteristic economic society.

On 26 December 2017, three regions including Shushan of Hefei, Xindu of Chengdu, and Rugao of Jiangsu signed contracts with CSISC and Home Big Data to specifically apply the results of big data to county governance, boosting the development of local big data applications and promoting the construction of Digital China.

The main component of Big Data Research Service Base is the "3 + 3" system, which refers to three key functional institutions and three affiliated institutions.

- Three key functional institutions—Research Center of National Big Data Laboratories in the Base, Specialty Big Data Industrial Park, and Service Cloud Platform of the Application of Big Data.
- Three affiliated institutions—the Innovation and Entrepreneurship Incubation Center of Big Data, the Practical Training Base for Talents of Big Data, and the Exhibition and Exchange Center of Big Data Results.

The base is constructed according to the basic principle of "Adapted and Customized to the Local" and the local characteristics. The three key functional institutions are necessary and can be focused and scaled adapting to the characteristics and needs of the actual region; while the three affiliated institutions can be built depending on the needs with customized selection.

The base adopts the model of joint construction by the CSISC, Home Big Data, and the local government, in which the local government and the CSISC are the joint competent units who jointly establish the executive management agency of the base to cooperatively promote the construction and management. Home Big Data, as a domestic authoritative big data research and service institution, is the implementation and operation service unit of the base, while the local government can also designate local enterprises to participate in the joint implementation and operating service.

The base focuses on the research and business services of big data technology for local pillar or characteristic industries, taking the integration of big data with local industries as the key goal in the application, to provide diversified big data services for local government, enterprises, and people's livelihood. The construction of the base can be varied on local characteristics and needs, adopting different strategies such as "research first", "application first", and "industry first" when implemented,

Sketch of the Development of Big Data in China

to optimize the investment of resources, the speed of progress, and the achievement of final results.

This model of joint construction of a big data research service base is of great value and significance to the development of the local economy and society.

1. **Digital Governance**. Through the construction of the base for the modernization of local governance system and governance capacity, develop comprehensive solutions for data management platform of local governance according to the local characteristics, and present a good demonstration for the construction of digital China, the national governance system, and the modernization of governance capacity.
2. **Emerging Economy**. Through the construction of the base, directly tap into the advantageous resources of local industries, develop new economic sectors, and obtain direct local economic benefit enhancement.
3. **Industrial Upgrading**. Through the base, upgrade the policy guidance of the park and the service output of the big data service platform, allowing big data to be integrated with local industries to promote industrial upgrading.
4. **Incubation of Innovation**. Through the base, upgrading the service functions and the research center of the park, accelerating the cooperation between industry, academia, and research and the incubation of innovations, thus promoting the development of local incubation of innovative enterprises.
5. **Service Upgrading**. The public services, government services, and corporate services provided by the big data service platform, provide the public with the services needed for life and work.

Big data research service base has both a standardized model design and can also fully reflect the flexibility and individual needs. For example, the customized digital town with local characteristics that accompanies the base model is a finishing touch to the construction of a new type of smart city. The co-construction and cultivation of the base is not only a driving innovation system and development model, but also a very valuable exploration and demonstration of the deep integration of government, industry, academia, and research in the development of big data applications in China.

5 The Brand-New Big Data Administration

The proposal to establish a Big Data Administration was first raised at the 5th Plenary Session of the 10th CPC Guangzhou Committee held in January 2014, which was supposed to coordinate and promote the collecting, sharing, and applying of information from government departments, and eliminate information silos to establish an open mechanism for public data. The proposal was soon supported by legislation.

A month later, the People's Government of Guangdong Province issued the Regulations on the Main Responsibilities and Staffing of the Economic and Information Commission of Guangdong Province, which explicitly mentions the establishment of the Guangdong Provincial Big Data Administration.

In May 2015, the People's Government of Guangzhou Municipality announced the "Sanding Scheme" issued by the Commission of Industry and Information Technology, the Commission of Commerce, and the State-owned Assets Supervision and Administration Commission. According to the proposal, the Commission of Industry and Information Technology will establish the Guangzhou Municipal Big Data Management Authority, which is responsible for researching, formulating, and organizing the implementation of big data strategies, planning and policy measures, guiding and promoting big data research and application, and organizing the formulation of standards and specifications for big data collection, management, opening, and application, with a total of nine functions. According to the regulations, there are three sections of the Guangzhou Municipal Government Affairs Data Administration—Standards Planning Section, Data Resources Section (Video Resources Management Section), and Information Systems Construction Section.

A few days later, Shenyang, Liaoning Province, also established a bureau-level big data management authority in Shenyang, with three divisions of Big Data Industry Department, the Standardization Department, and the Data Resource Department. It took no longer than 3 months for the authority from the construction to operation.

On 1 September 2015, Chengdu City approved the establishment of a big data management authority under the Municipal Commission of Economic and Information Technology. This is another new functional authority constructed by the local government, following Guangdong and Liaoning provinces. Most of these fledgling authorities are directly under the regulation of local government and are not ranked in low the hierarchy. Its arising will create synergy in the coordinated planning of government data resources and social data resources, as well as in the improvement of basic information resources and the construction of important information resources, which will be conducive to the early formation of cyberspace where people and machines interact and everything is connected in China.

Many people are curious about the establishment of a Big Data Management Authority in line with the trend. What it will do? How does it differ from the existing statistics bureau? Table 1 is the related information of 15 local big data management authorities that have been established in China and their main functions, which are compiled from publicly available information to assist in understanding.

Sketch of the Development of Big Data in China

Table 1 Local big data management authorities and their main functions

Name of institution	Affiliated departments	Established time	Level	Main functions
Guangdong Provincial Big Data Administration	Economic and Information Technology Commission of Guangdong Province	February 2014	Department-level	1. Research, formulate, and organize the implementation of the strategies, planning, and policy measures of big data, guide and promote the research and application of big data 2. Organize the development of big data collection, management, openness, application, and other standards and norms 3. Promote the establishment and applications development of the formation mechanism of big data for the whole society 4. Undertake the comprehensive work of enterprise situation, responsible for enterprise data collection and storage 5. Organize the planning and implementation of e-government construction 6. Organize and coordinate the sharing of government information resources 7. Organize and coordinate the construction of major provincial e-government projects; organize and coordinate the construction of one-stop e-government services such as an online service platform 8. Responsible for coordinating the construction and management of government information network systems and government data centers 9. Coordinate the construction of the information security system, undertake the protection of information security level, emergency coordination, and related work of digital authentication

(continued)

Table 1 (continued)

Name of institution	Affiliated departments	Established time	Level	Main functions
Guangzhou Municipal Big Data Administration	Industry and Information Technology Commission of Guangzhou Municipality	May 2015	Division-level	1. Research, formulate, and organize the implementation of the strategies, planning, and policy measures of big data, guide and promote the research and application of big data 2. Coordinate and organize the implementation of national and local data technology standards, and study and develop the basic information resources sharing and exchange directory, technical specifications and scope 3. Organize the development of big data collection, management, openness, application, and other standards and norms 4. Develop and organize the implementation of smart city operating management specifications and component standards 5. Promote the establishment and applications development of the formation mechanism of big data for the whole society 6. Responsible for coordinating the planning and construction of the industrial database, the establishment of enterprise energy consumption, environmental protection, safety production monitoring indicators, and other databases to support the operation of the public information platform with the integration of informatization and industrialization 7. Responsible for enterprises' data information collection, statistics, analysis, reporting, and other comprehensive work on the general enterprises' situation 8. Organize the construction of the public information platform and industrial big data platform with the integration of informatization and industrialization; coordinate the construction of intelligent video systems for urban management; promote the integration and sharing of video resources and comprehensive applications 9. Undertake the promotion and application of supercomputing and cloud computing technology platforms in Guangzhou

(continued)

Table 1 (continued)

Name of institution	Affiliated departments	Established time	Level	Main functions
Shenyang Municipal Big Data Administration	Economic and Information Technology Commission of Shenyang Municipality	June 2015	Bureau-level	1. Responsible for organizing and developing the overall planning and implementation of smart Shenyang; research and develop big data strategies, planning, and related policies 2. Organize and develop the standard system and assessment system of big data; coordinate and promote the construction of a big database for the whole society; organize and develop the standard disciplines on big data collection, management, opening, trading, and application 3. Guide the development of big data industry 4. Research, develop, and organize the city's overall planning and implementation plan of municipal e-government construction 5. Organize and coordinate the sharing of government information resources 6. Coordinate the construction of an information security guarantee system
Chengdu Municipal Big Data Administration	Economic and Information Technology Commission of Chengdu Municipality	September 2015	Division-level	1. Responsible for the development of the city's big data strategies, planning, and policy measures and organizing the implementation 2. Promote the standard disciplines of information data collection, management, openness, and application 3. Promote the interconnection, interaction, and resource sharing of information data resources and infrastructure construction 4. Develop, and organize the city's overall planning and implementation plan of municipal e-government construction; lead the audit of e-government project 5. Promote the integration of existing information systems in the e-government from external networks; organize and coordinate the construction of the city's information security system 6. Undertake the daily work of the Office of the Leading Team of Municipal Informatization

(continued)

Table 1 (continued)

44

Name of institution	Affiliated departments	Established time	Level	Main functions
Lanzhou Municipal Big Data Social Service Administration	People's Government of Lanzhou Municipality	September 2015	Sub-department-level	1. Implement the major decisions and plans of the central, provincial, and municipal parties on information technology construction 2. Research, develop, and organize the implementation of the city's informatization social services, smart city construction, and other planning and policy measures of the development of big data; in conjunction with relevant departments to study and implement the policy planning towards information industry 3. Responsible for coordinating and promoting the construction of a big database for the whole society; organizing and developing the standard disciplines on big data collection, management, opening, trading, and application 4. Responsible for the informatization construction and management of the city's social services; responsible for the review of proposals and programs on overall and departmental informatization construction across the county and district 5. Responsible for the audit, the use of funds, operating guidance, and result assessment of major projects towards informatization across the city 6. Responsible for coordinating and promoting the integration and sharing of information resources across the city 7. Responsible for the guidance, supervision, and assessment of big data industry, three-dimensional digital social services management, and information industry promotion 8. Complete other work assigned by the municipal party committee and the municipal government

(continued)

Table 1 (continued)

Name of institution	Affiliated departments	Established time	Level	Main functions
Guizhou Provincial Big Data Development Administration	People's Government of Guizhou Province	October 2015	Department-level	1. Research, develop, and organize the implementation of big data strategies, planning, and policy measures; guide and promote the research and application of big data 2. Organize the development of the standard system and assessment system of big data; develop the standard specification of collection, management, and opening and application of big data 3. Responsible for the management of the information industry led by big data, coordinating and promoting the development and application of big data in various fields of social economy 4. Organize and implement the planning of e-government construction project Internet + Action; organize and coordinate the construction of municipal major e-government projects and Internet + Application Projects 5. Responsible for information infrastructure planning, coordination, management, and supervision 6. Responsible for coordinating the construction and management of government information network systems and government data centers; promoting the sharing and opening of government data resources 7. Coordinate the construction of an information security assurance system and undertake the work related to the protection of information security level, emergency coordination, and digital authentication 8. Complete other work assigned by the municipal party committee and the municipal government
Baoshan Municipal Big Data Administration	Industry and Information Technology Commission of Baoshan Municipality	November 2015	Division-level	1. Promote the integration, opening, and utilization of data resources 2. Deepen the Internet + and Citizen Beneficial Project 3. Promote the development of Baoshan International Data Service Industrial Park

(continued)

Table 1 (continued)

Name of institution	Affiliated departments	Established time	Level	Main functions
Huangshi Municipal Big Data Administration	Economic and Technology Information Commission of Huangshi Municipality	November 2015	Sub-division-level	1. Research, develop, and organize the implementation of big data strategies, planning, and policy measures; guide and promote the research and application of big data 2. Organize and develop the standard specification of collection, management, and opening. and application of big data 3. Promote the development and application of big data in various fields of social economy 4. Undertake comprehensive work of enterprise situation; responsible for enterprise data collection and storage 5. Responsible for coordinating the construction and management of government information network systems and government data centers
Xianyang Municipal Big Data Administration	People's Government of Xianyang Municipality	July 2016	Department-level	1. Responsible for the organization and leadership of the city's information sharing work; coordinating and solving major issues related to government information sharing 2. Research, develop, and organize the implementation of municipal big data strategies, planning, and policy measures; guide and promote the research and application of big data 3. Organize the city and develop the standard specification of collection, organization, reporting, maintenance, and updating of big data 4. Establish a municipal unified data service center and information sharing mechanism 5. Implement dynamic management of information sharing to ensure the accuracy and effectiveness of shared information 6. Coordinate the security management of information resources and undertake the protection of information security level, emergency coordination, and digital authentication

(continued)

Sketch of the Development of Big Data in China

Table 1 (continued)

Name of institution	Affiliated departments	Established time	Level	Main functions
Yinchuan Municipal Big Data Administration	People's Government of Yinchuan Municipality	November 2016	Division-level	1. Research and organize the implementation of Smart Yinchuan construction and big data strategies, planning, and policy measures; guide and promote the construction of Smart Yinchuan and the research and application of big data; coordinate the interconnection, interaction, and information sharing of information resources 2. Coordinate and organize the implementation of national and local data technology standards; study the directory, technical specifications, and scope of basic information resource sharing and exchange 3. Organize and implement the standard system and assessment system of big data; promote the establishment, development, and application of big data formation mechanism 4. Organize and implement the standards and specifications for big data collection, management, opening, trading, and application 5. Organize and implement the operating management specification and component standards of Smart Yinchuan 6. Coordinate the construction of the city's information security guarantee system

(continued)

Table 1 (continued)

Name of institution	Affiliated departments	Established time	Level	Main functions
Hangzhou Municipal Data Resource Administration	People's Government of Hangzhou Municipality	January 2017	Bureau-level	1. Implementation of national and provincial policies, laws, and regulations on data resource management; entrusted to draft the city's relevant local laws and regulations, draft regulations, and normative documents, and organize the implementation after consideration and adoption; develop the city's data resources development strategy, development planning, annual plans, policy measures, and evaluation system, and organize the implementation 2. Organize and implement the national and local data technology standards; research and develop the standards and norms of the city's data resources collection, storage, use, opening, and sharing, and supervise the implementation 3. Responsible for the construction and management of the city's government data and public data platform; organize and coordinate the directory development, collection, and management, organization and utilization, sharing and opening of the city's government data and public data resources; promote the application of data resources in the field of government management and social governance; organize and implement major plans of the city such as the "Data Brain" and others 4. Guide the data development and utilization of the city's economic and social data in various fields; promote the development of the city's big data industry 5. Guide the construction of the city's data security system; organize and implement the city's government data and public data security 6. Organize and coordinate the construction, management, and performance evaluation of smart e-government projects and data resource infrastructure of the municipal government system 7. Responsible for the construction and management of the "China Hangzhou" government portal; guide the e-government construction and portal construction of local districts, counties (cities) government, and municipal units; undertake the construction and management of the municipal government information open data resource platform and service security 8. Responsible for foreign exchanges and cooperation of the city's data resources and government informatization; in conjunction with relevant departments to do a good job of the city's big data, government informatization personnel training, and other work 9. Undertake other matters assigned by the municipal government

(continued)

Table 1 (continued)

Name of institution	Affiliated departments	Established time	Level	Main functions
Inner Mongolia Big Data Development Administration	People's Government Inner Mongolia Autonomous Region	January 2017	Department-level	Comprehensively implement the spirit of the 18th CPC National Congress and the 3rd, 4th, 5th, and 6th Plenary Sessions of the 18th CPC Central Committee. Thoroughly implement the spirit of President XI Jinping's important speeches of new ideas and new strategies for governing the country. Conscientiously implement the decision-making department of the Party committee and government of the autonomous region, giving full play to the supporting role of big data for the synergistic development of "Standardization of Branch Setups, Normalization of Organizational Life, Refinement of Management Services, Systematization of Work System, and Standardization of Workplace Construction", the pioneering role for the construction of "Seven Networks of Modern Water, Comprehensive 3D Transportation, Modern Logistics, Energy Security, Municipal Public Facilities, New Infrastructure, and Rural Technology Facilities", and the catalytic role for the development of "Seven Industries of New Energy, New Materials, Energy Conservation, and Environmental Protection, High-end Equipment, Big Data, and Cloud Computing, Biotechnology, and Inner Mongolian Chinese Medicine". We will take the construction of the national big data comprehensive pilot area as a grip, insist on taking the foundation of facilities, security as a prerequisite, resources as the root, and application as the core, to strengthen the information infrastructure, promote the integration, sharing, and opening of government data resources and innovative applications, vigorously develop big data industry, and promote the transformation and upgrading of the region's economy and society
Kunming Municipal Big Data Administration	People's Government of Kunming Municipality	March 2017	Sub-department-level	1. In accordance with national and provincial requirements to develop a big data standard system and assessment system; organize and implement the collection, management, open, trading, application, and other related work of big data 2. Coordinate and promote the wild application of big data in various fields of social economy 3. Coordinate the overall promotion of the construction of smart cities

(continued)

Table 1 (continued)

Name of institution	Affiliated departments	Established time	Level	Main functions
Shenzhen Longgang Big Data Administration	People's Government of Longgang District	January 2018	Division-level	1. Promote the construction of informatization and the development and application of big data in the region to create Digital Longgang 2. Promote the construction of Longgang smart city; optimize and improve digital infrastructure; integrate government data and social data resources; ensure data security; form a sharing and opening system 3. Promote the construction of the leading data area; utilize big data to enhance the modernization of government governance; promote the innovation and development of big data technology industry; protect and improve people's livelihood; promote the intersection and integration of information technology and the economy and society
Xuzhou Municipal Big Data Administration	People's Government of Xuzhou Municipality	January 2018	Division-level	1. Responsible for the development and implementation of the city's big data strategic planning and policy measures 2. Develop the standard system of big data; develop municipal standards and specifications for collection, management, openness, application, and transaction of big data; coordinate and promote the development and application of big data in various fields of social economy 3. Promote the sharing and opening of government data resources 4. Responsible for the construction of the city's big data information security technology assurance system

Managing Changes in the Era of Big Data

Qing Jiang

1 Changes of Big Data Facing Organizations

Lack of cohesion is the biggest obstacle to organizational management. The cohesion that was not supposed to be quantified is measured by big data, for managers to know what is the core cohesion of the team and thus creating a centripetal force for effective management. From my years of research in the applications of big data, I rather think "the Internet has given rise to the advent of the era of information technology, and information technology has made the era of big data." Today, one of the points I would like to elaborate on is that the era of big data has arrived, and leaders of government organizations must follow the trend and welcome the arrival of this new era with a positive mindset.

Ben Waber, CEO of Sociometric Solutions, a famous US company, begins his book Big Data Management with the following words: "What if I told you that a change in work breaks would make employees more productive, and that how would you feel if I told you that changing the size of the dining table would make employees more productive and that re-sizing the table could be an important decision for the company? These are detailing that traditional human resource theory has never looked at because they are difficult to quantify. However, taking an extra 10-min break, changing the position of the water cooler, and eating lunches with colleagues more often are some of the things that have previously gone unnoticed by business managers and are part of what makes a team more cohesive and committed."

As everyone knows, the main role of the Internet is the dissemination and sharing of information, and its most important organizational format was to create a static website. Since 2004, social media such as Facebook, Twitter, Weibo, and WeChat have been created, marking the advent of the new era of the Internet. In this new era, the Internet has become a vehicle for real-time interaction and communication.

Q. Jiang (✉)
Zhongnan University of Economics and Law, Wuhan, Hubei, China
e-mail: 18712919290@163.com

© China Renmin University Press 2023
Q. Jiang (ed.), *Digital China: Big Data and Government Managerial Decision*,
https://doi.org/10.1007/978-981-19-9715-0_4

Organizational change in the era of big data will be reflected in the following areas.

1. **Cross-Functional and Cross-Departmental Data Flows**. Effective organizations should allocate information and decision-making authority to different departments. In the era of big data, flexible organizational structures and maximum cross-functional cooperation are the necessary basis for organizational development. Leaders need to provide departmental managers with the right experts in data and technology. At the same time, IT planning and operations should be given high priority by leaders, as a sound IT structure of enterprises can also help solve the problem of information silos.

2. **Clearly Define Data Needs**. Some argue that the importance of leaders having a good experience, intuition, and vision will be less decisive in the era of big data. In this era, leaders who can identify opportunities, think creatively, and convince employees to invest in new ideas are required instead, thus these leaders must be able to adapt to the era of many decisions when big data itself is also one of those adaptations.

3. **Managing Data Technicians**. Data technicians will be particularly valuable in the era of big data, the most important of which are scientists who can process big data. Statistical skills are essential for data scientists, but more important are their skills in cleaning and organizing large data, as data formats in the era of big data are often unstructured. The best data scientist must not only know the language of the business but also have a complex knowledge base and even experience in the field, only then can they help leaders understand the challenges faced by their organizations from a data perspective.

4. **Real-time User Customization**. In the era of big data, personalized service will overturn the traditional model and become the direction and driving force for the development of society in the future. Big data provides ample space for sustainable development for personalized applications, data that flows based on cross-fertilization, holographic individual behavior, and preference data, etc. Through research and analysis of these data, society management and business applications in the future can precisely tap into the different interests and preferences of each individual and thus provide them with personalized services.[1]

5. **Data-Based Operations and Decision-Making**. Big data can be used to further leverage algorithms and machine analysis to improve operational and decision-making efficiency. Big Data has given rise to new information-driven management models and plays a backbone role in the organizational value chain. Data-driven decision-making methods can validate assumptions and analyze results, which will lead to decision guides and operational changes.

In addition, big data will have a profound impact on work performance and human resources. The era of information technology has provided leaders with a vast and deep ground for applications of big data. As long as the leaders can change their

---

[1] SU Meng. Business Changes in the Era of Big Data [J]. China Computers & Communication 2012 (11).

Managing Changes in the Era of Big Data

minds of thinking as soon as possible, learn further and make full and effective use of big data, they will be able to seize the opportunities and win the competition in the fast-changing globalized economic environment.

2 Embarrassment Encountered in Market Research

Market research is an important tool for managers to understand the current state and development trends of the market, and to make decisions through analysis. However, due to regional scope and technical conditions, the number of samples obtained from the survey is limited, so the results of the analysis are inevitably wrong. In this case, the decisions made are often contrary to expectations, which could even cause significant losses to the organization. However, many leaders have been accustomed to listening to experts or making decisions based on experience, market research for decision making in many organizations is in an extremely awkward position.

In the late 1970s, the rise of PepsiCo forced the long-established Coca-Cola to respond to the fierce challenge of this "rising star". In fact, the strategic intentions of PepsiCo were so obvious to impact the Coca-Cola market through a large number of energetic and stylish advertisements. Firstly, PepsiCo targeted the largest consumer group in the beverage market—young people—with an advertising campaign based on themes such as adventure, passion, youth, and ideals, thus winning the youngster. At the same time, PepsiCo launched a creative campaign "taste test", in which subjects were given a taste of two unmarked cola flavors without any label mark or sign, and the results showed that more than 80% of people considered that Pepsi tasted better than Coca-Cola.

In response, Coca-Cola set out to find out why Coca-Cola was not as good as Pepsi by launching a market research initiative with the code name "Project Kansas". It has reached out to 10 major cities and interviewed around 2,000 consumers, asking their opinions on the new Coke flavor with questions such as "Would you like to try the new drink?" "Would you be satisfied if Coca-Cola became gentler in flavor?" The results of the research showed that consumers were willing to try new flavors of Coke, which reinforced the belief of Coca-Cola's decision makers that the 99-year-old Coca-Cola formula is no longer suitable for needs of today's consumers. Hence, with confidence, Coca-Cola set out to develop a new flavor of coke.

Soon, Coca-Cola brought in a new sample with a gentler, sweeter taste with less foam than the original coke. They have also made a huge advertisement campaign before its official launch to the market. On 23 April 1985, Coca-Cola held a press conference at New York's Lincoln Center to celebrate the launch of the new Coke with over 200 journalists from newspaper agencies, magazines, and TVs attending, which created a huge buzz.

At first, the new coke was sold well, with 150 million people trying it out. However, the new formula was not acceptable to everyone, which was not because of the taste, but the "change". The original Coca-Cola consumers were rejected or even got angered. They felt that the 99-year-old secret formula of Coca-Cola represents a traditional American spirit, and the abandonment of the traditional formula is a betrayal. In Seattle, a group of traditional Coca-Cola coke loyalists has formed the organization "Old American Coke Drinkers" and prepared to launch a nationwide "New Coke Boycott". In Los Angeles, some consumers have even threatened that they "will never buy Coca-Cola again if they introduce new coke."

At the same time, the old flavor of Coca-Cola was in short supply and prices were continuously increasing. Every day, the company received a flood of letters and phone calls from angry consumers, and the decision-makers had to change their decisions by the pressure to boycott the new coke. In response, they conducted another survey of customers' intentions, which showed that 30% of people preferred the new flavor, while 60% categorically rejected it. As a result, Coca-Cola had to resume production of the traditional formula while retaining the capacity for the new coke.[2]

Within less than three months in 1985, Coca-Cola spent $4 million on a new product that, despite two years of previous research, ended in failure. The birth and popularity of several contemporary Internet brands, such as Xiaomi and Three Squirrels, on the other hand, has completely overturned the traditional approach to market research. We should be grateful for the advent of the Internet and the era of big data. In the past, without the Internet, the market research practiced by enterprises has inevitable limitations and bias. Decision makers cannot accurately grasp and analyze the entire market, which leads to mistakes in the decisions. Only with enough user data accumulated by the Internet, can we analyze the preferences and needs of the majority of people for a certain product, so that we can achieve true precision marketing and avoid more mistakes in decision making. When making product adjustment decisions in response to new customer demand, business managers should also pay attention to interactive communication with old customers to cultivate their loyalty to the brand, establishing a continuous and effective marketing management channel for the company. And this vision cannot be achieved without the support of the Internet and big data.

3 We Are All Managers

Anyone with experience in management knows that there is a limit to the number of direct subordinates, so when the organization grows, it has to be decentralized to form management teams and adopt indirect management.

The world-famous "Amoeba Operating" is a successful example of INAMORI Kazuo's implementation of indirect management to achieve full participation in management, which has enabled Kyocera, one of the world's top 500 companies, to operate and lead for more than 50 years, creating fabulous results. However, achieving full participation in management requires certain conditions. Firstly, management must be based on mutual trust, and without this, important operational information cannot be made available to employees. Managers must be aware that employees shall

---

[2] Marketing [J]. Seed World. 2005 (1).

not be seen simply as tools but as part of the management community. Secondly, the Amoeba Operating is a system that allows on-site employees to make judgments and take action based on data. Therefore, it is essential that data is provided in a timely manner. Employees will be seriously discouraged if wait too long until everything is irreparable or even hold accountable to them. Thirdly, senior management needs to be problem-solving oriented, together with the Amoeba team, conducting on-site training based on real-life cases for employees. Employees who lack a certain level of knowledge are not able to identify problems and find reasonable solutions based on operational data. Therefore, this training is essential, especially in the early stages when introducing staff. It is impossible to achieve truly full participation in management by leaving the operation to the field staff. Fourthly, each member of the Amoeba must have a holistic view of the company as a whole, otherwise, there is a risk that they will do whatever it takes to achieve their goals, causing internal conflict instead.[3]

Alex Pentland, director of MIT's Human Dynamics Lab, wearables pioneer, and expert on big data stated in his article New Science to Shape the Great Team published in the Harvard Business Review in April 2012, "to liberate oneself from the management of organizational charts, one needs to abandon the approach of relying on individual talents, but manage organizations with better collective intelligence through shaping patterns of interaction."

In 2009, the Defense Advanced Research Projects Agency (DARPA) organized a "Red Balloon Challenge". Ten red balloons were set across the United States and whomever the organization or individual could find the coordinates of all within the shortest time would win the $40,000 prize. It was a super challenging task, one senior analyst at the National Geospatial-Intelligence Agency (NGA) called it "insurmountable by traditional intelligence gathering methods".

More than 4,000 teams in total in the United States participated in the competition. In the end, the team led by Alex Pentland became the first to complete, marking all coordinates in only 8 h, 52 min, and 41 s. In fact, they knew about the activity only a few days before the balloons were placed, but within a few hours they had assembled a team up to 5,000 people. Each member of the team notified an average of 400 friends, making a total of around 2 million people helping the Pentland team to complete the challenge!

So, what exactly did the Pentland team do to mobilize such a large group of 2 million people? Obviously, traditional organizational management methods were difficult to navigate and Pentland was actually using a social network incentive strategy. This was done by rewarding not only those who correctly informed the team of the balloon's location but also those who successfully introduced the person who found the balloon to the team. Pentland divided the $40,000 into 10 equal shares of 10 balloons (each balloon corresponds to an award of $4,000) and promised $2,000 to the first person who told the team where the balloons were located, $500 to the

---

[3] RAN Mengshun. A Study on the Application of Amoeba Operating Approach in Company S [D]. Suzhou: Soochow University. 2014.

person who introduced that first person to the team, $250 to the person who introduced the previous person, and so on.

The longest communication chain among the team for finding the balloon was 15 people! In addition, one-third of the Twitter records that helped spread the team's message came from outside the United States. Even though those who don't live in the continental US cannot find the balloon directly, it's entirely possible for them to use other ways to spread the team's message.

So how did Pentland discover the mystery of social network motivation? Here we have to mention Pentland's mastery of big data technology. Before the advent of big data, it was impossible to capture real-time data on how team members in an organization communicate with each other (including information such as tone of voice, body language, talking subject, and communication period). Pentland's genius was to develop a device called the "social relationship meter". The device consists of a location sensor, an accelerometer to record body language, a proximity sensor to determine who is nearby, and a microphone to record if someone is talking (and to avoid invading privacy, the device does not record voice content or video).

Pentland has studied innovation teams, post-operative care units in hospitals, customer service teams in banks, back-office support, and call center teams with sociometric signage, to uncover the mysteries of human interaction and collective intelligence within organizations that lie behind big data. The study found that the productivity of a team can be accurately predicted by measuring the interaction patterns of that team. Groups with a more balanced distribution of discourse shifts have higher collective intelligence than groups in which a few authoritative individuals dominate the conversation. The centralization, control, and prevention models common to traditional management methods stifle the potential of collective intelligence and the overall effectiveness of management. Pentland was able to unlock the collective intelligence of 2 million people to its fullest potential in the Red Balloon Challenge only by revealing this mystery.

Thus, it can be seen that there is no hierarchy of managers. The position determines the management role and everyone is a manager in their position. The above two examples are worthy of reference for all leaders facing transformation.

Thoughtless Decision-Making Versus Big Data Thinking

Qing Jiang

1 Criticism of Captain's Call

The "Captain's Call" refers to the phenomenon that making decisions relying on one's own unreasonable imagination without investigation or research. Organizational decision-making is often a big issue directly related to its survival, as one mistake can have serious consequences. However, in recent years, many senior managers are accustomed to making captain's calls, which are mainly manifested in the process of decision-making that is often leadership-oriented. Firstly, the leader has a judgment of their own, then use the democratic way to identify people who share the same opinion, or request subordinates to find some data to support and prove the correctness of this judgment. The subordinates with concerns of personal interests or another psychological impact will, of course, tend to rack their brains to find some supporting data from any possible dimensions to prove that the leader's point of view is correct. As a result, some projects or goals set by the leaders themselves become the direction for the normal person to follow, even if they may be wrong. This results in four embarrassing phenomena: making captain's calls, giving guarantee without throughout consideration, regretting after failure, and leaving without taking responsibility.

Decision-making based on intuition and experience has been an epidemic for some leaders for quite a long time, ignoring that the real world is complex and the external environment is ever-changing. The risk of making such captain's call decisions is high and people are likely to pay a serious price for them. As we all know, people are not machines, and no matter how good the intentions are, the decisions are bound to be influenced by personal consciousness and many decisions will turn out to be wrong. For example, the leaders of a county in the north used their own "ideas" replacing the urban construction department, and tried to build

---

Q. Jiang (✉)
Zhongnan University of Economics and Law, Wuhan, Hubei, China
e-mail: 18712919290@163.com

© China Renmin University Press 2023
Q. Jiang (ed.), *Digital China: Big Data and Government Managerial Decision*,
https://doi.org/10.1007/978-981-19-9715-0_5

green streets with seven different species of trees including palm, spruce, Osmanthus fragrans, bamboo, willow, boxwood, and paulownia to "reflect the southern scenery". However, the survival rate of those species, which are only suitable for the southern environment, is extremely low in the north, resulting in the repeat in their planting and dying. In order to maintain these "vanity projects", the county spent up to 13 million yuan of state funds on poverty alleviation. It is no coincidence that in the 1990s, it became popular to build development zones all over the country, and many local governments from provinces to cities and towns ignored their actual situation and launched projects, when the total development zones of various sizes and names reach 10,000. As a result, many of these development zones have been deserted, and huge amounts of money have been lost.[1] Shenyang Green Island Sports Center, which had the largest indoor football field in Asia and cost around 800 million yuan to build, was demolished after less than 10 years of use. According to media reports, the reason for the demolition was its low usage. This is also a typical case of poor government decision-making: for one thing, the Green Island Sports Center was located in a suburban area with poor transportation; for another, it was in competition with large stadiums such as the Olympic Sports Center and Tiexi Stadium, resulting in a waste of resources. In recent years, a variety of signature town projects have been brought up everywhere, and some imitating "works" have been made by some leaders as the captain's call. In return for the phenomenon that "thousands of towns looking the same", the National Development and Reform Commission, the Ministry of Land and Resources, the Ministry of Environmental Protection, and others jointly issued the "Opinions on Standardizing the Construction of Signature Towns and Signature Counties", which ushered in new challenges and opportunities for the development of signature towns as an important way to implement the country's new urbanization.

Some government leaders are keen to make captain's call, while some business leaders are repeating the same way. The key project of China's pharmaceutical investment Henan Zhongyuan pharmaceutical plant covers an area of 1300 mu, with a total investment of 1.8 billion yuan, which has been shut down and closed without starting production. The reason for this is that without detailed investigation, one of the key technologies of the project was introduced by a small Swiss company with only twenty-odd employees.

In practical analysis, most the wrong decisions were made because the decision maker was too emotional, the captain's call happened either by one's arbitrary and unconventionality or by formalistic democracy from the collective. In this process, we are influenced by emotions, which can limit the direction of our choices without allowing more possibilities into our vision. Decisions are made without scientific analysis beforehand and procedural constraints, just simply by making the captain's call to get inspiration, which is the most important cause of instability in organizational development. If the inspiration is right, the organization will get one chance to develop; if the inspiration is wrong, the organization will stagnate and even fall into difficulties. Some leaders have also put their hopes on "research and democratic decision-making" and "hiring management consultants and experts to prove

---

[1] Shuixiangke. The Unbearable Failure of Decision Making [J]. Prosecutorial View, 2006(23).

the conclusions". In fact, the so-called research and democratic decision-making are still around the leadership. At best, it is some research in the local community to find pieces of evidence that support the main leadership, and find a few departments or representatives to hold a forum where people can vote on it. Let us not mention the effectiveness of such small-scale research and democratic decision-making, even the time wasted is not affordable.

It is easy to see from the above examples that most of these negative consequences are the result of the "epidemic" of leaders blindly believing in "captain's calls". Decision-making is a choice of how to allocate resources, seeking to maximize the organization's progress towards its development goals. It can only be achieved if the organization maximizes external opportunities in the way it chooses to allocate resources and makes the most of internal resources without exceeding its limits. The relationship between the two is not something that can be grasped by gambling on intuitive inspiration. If any organization is to achieve sustainable development, it must move away from inspirational decision-making based on intuition and replace it with procedural decision-making based on scientific analysis.

2 Big Data Thinking Disrupts Tradition

McKinsey estimated that if companies or organizations could take full advantage of big data analytic, it could generate $300 billion in revenue annually for the US healthcare industry, €250 billion for the European public sector, and possibly more than 60% profit for the retail industry. The new meaning that big data gives to thinking and decision-making is driving the global economy to rethink business models and consider changes in decision-making patterns, and the huge demand for big data is driving governments and businesses to move more rapidly towards big data practices.

In 2011, Watson, the world's most popular supercomputing system, won the *Jeopardy* quiz competition against human competitors. At the time, industry experts from all over the world made three major predictions about the future of Watson's applications: improving access to healthcare, helping with major financial decisions, and improving and optimizing customer service processes.

In recent years, these three applications have been accompanied by innovative practices that have moved from prediction to reality. Watson as a "physician's assistant" is gaining ground in the healthcare industry: on Wall Street, Watson is providing cloud services to Citibank to manage and predict portfolio investment risk; Nielsen, Royal Bank of Canada, and others are using Watson as a "cognitive assistant" to quickly organize vast amounts of user data to help customer service specialists interact with customers around the world.

In the field of neonatal care, global experts have been working for years to better integrate medical diagnostics with science and technology to help healthcare professionals improve the survival rate of babies, especially premature babies, during the most critical 24 h of giving birth by using cutting-edge big data processing and analysis. The neonatal intensive care specialists of the University of Ontario Institute of

Technology are required to use IBM Big Data Software to analyze premature babies or sensors and detection devices on them, with more than 1,000 unique pieces of vital information per second, which allows the caregivers to identify life-threatening factors in real time, as well as predict symptoms that are likely to occur in the next 24 h, so timely action can be conducted.

The first thing that needs to be addressed to make data valuable is the thinking of the leaders. If leaders cannot be aware of the value of data, the normal personnel, no matter how pursuing and enthusiastic, will not be able to reflect the value of data, which is also inextricably linked to the current status of our social and economic development. Today, with the advent of the Internet, cloud computing, and the era of big data, the collection and application of big data is no longer a problem. Leaders, supported by a team of big data talents, can grasp the dynamic information, intelligent analysis of the current situation, and think-tank references they need in real-time through computers.

As a data developer and operator, ZHENG Jie, Chairman and General Manager of China Mobile Communications Group Co., Ltd (Anhui), said in the context of his own work and analysis: "Once the thinking has shifted, data can be cleverly used to inspire new products and new services. Big data-based solutions are an important measure of industrial upgrading and productivity improvement. Discovering value from data and generating wealth is what makes big data most attractive."

We know that the core objective of big data technology is to extract the hidden patterns from the data with its vast structure, huge volume, and diverse characteristics so that the data can be used to its maximum value. We have to understand the fact that big data has driven changes in the way we work and live now, and that people have to use big data thinking to face the era of big data if they want to make use of it.

Human society has evolved from the age of information technology to the age of digital technology, and the biggest change is the need for everyone to change their way of thinking. People are the most important element in the governance and development of a country, and big data thinking is essential to the progress and development of society as a whole. Without a shift in thinking, even doing traditional things using new technology will be difficult to achieve. Whether it is in government, finance or industry, before the application of big data there must be a new way of thinking and a new model first.

From bringing a more exciting watching experience to the Wimbledon Championships to giving Watson the power of forward-thinking cognitive computing, the power of big data analysis is changing the world at a rapid pace. There is a huge amount of data being created every moment, which means that there is also a huge amount of value to be explored. The significance of big data analysis has long since transcended the playing field and the business world of competition. In the new world of big data, the whole society is a laboratory of innovation and discovery, in which you and I are creators, explorers, and change-makers. As saying thirty percent of technology and seventy percent of data, the one who can utilize the data will win the world. Changing the way of thinking is crucial while making the best use of big data thinking, using data to prove, and using big data to support decision-making has become a trend that modern leaders must face, and organizations and companies are even more a leader in this change. The era of big data has arrived, are you ready?

3 Public Opinion Monitoring and Data Analysis

Public opinion (i.e., social sentiment and public reaction) is a subjective reflection of social reality by the public within a certain period of time and within a certain scope, which is a comprehensive reflection of the thoughts, psychology, emotions, will, and demands of a group of people. In its superficial sense, it should include both social sentiment (objective) and public reaction (subjective). Various parties mostly emphasize its subjective nature when using this concept, while its essence refers to public sentiment and public reaction.[2] The pure concept of public reaction should be the distribution of public opinion on specific things, but most public surveys conducted on the state of social development nowadays already have the characteristics of social attitude surveys, so social sentiment and public reaction can also be defined as public reaction with the meaning of social attitude.

Public opinion on the Internet is a manifestation of social opinion, specifically referring to the dissemination through the Internet of highly influential and tendentious remarks and views held by the public on certain hot and focal issues in real life. In the online environment, the main sources of public opinion information include news commentaries, BBS, blogs, RSS, and other media.[3] Nowadays, online public opinion data is increasingly characterized by big data. In modern society, values are pluralistic in expression, and the exchange of various views is both convergent and conflicting. The pluralism of public opinion is manifested in diversity and variability, while the development of the Internet, especially the mobile Internet, has made online public opinion change exceptionally fast. Public opinion monitoring, on the other hand, integrates Internet information collection technology and intelligent information processing technology to automatically capture, automatically classify, and cluster, themes detect, and topics focus on the vast information on the Internet, in order to realize users' information needs and form analysis results such as reports and charts, so as to provide analytical bases for comprehensively grasping the public's thought dynamics and making correct public opinion guidance, crisis attack, and scientific decision-making.

The openness of the Internet enables a huge number of Internet users and various social groups to express their views online easily and quickly, which makes the

---

[2] WEN Shi. Several Channels for Broadening the Work of Reflecting Social Sentiment and Public [N]. Xilingol Daily, 2010-06-25.

[3] LIU Zhiming, LIU Lv. Identification and Analysis on Public Opinion of Key Opinion Leaders in Weibo [J]. Systems Engineering, 2011, 29(6):8–16.

volume of data on Internet public opinion grow rapidly. The development of multimedia has made the data form of Internet public opinion showing a diversity of characteristics including text, pictures, audio, and video.[4]

Leaders with their powers and responsibilities, always have to make important decisions directly related to social governance, economic development, and people's welfare. How exactly should the leaders have great ingenuity and unique wisdom, then seize the essence and uncover the root cause of the confusing information to avoid mistakes in decision-making? To utilize big data to assist leadership decision-making, it is necessary to integrate data sources for collection, and the five-step workflow of "collection, storage, management, research, utilization" on the home page can actually be explained as data collection, storage, and calculation, data management, research and analysis, and application of results respectively. The most valuable part of decision analysis is the association of data attributes. When the volume is large, traditional analysis software cannot be completed well, and even the data set cannot be opened. When data attributes are linked together, a more comprehensive analysis can be carried out, making the data more in-depth and informative. Integrating multiple structured data (including structured data, unstructured data, and semi-structured data, or divided into online data and offline data) from multiple data sources is very difficult which cannot be solved simply in one paper. From the perspective of technique, questions including whether the data source data is credible (such as whether the format and content of the data can be read and used by users), and thus whether it can facilitate in-depth processing and analysis of the data, as well as whether the accuracy, completeness, timeliness, and validity of the data can fulfill the requirement of application are difficult to solve quickly, while there are many other factors should be considered, which cannot be done in one day. However, the utilization of big data for decision support is beginning to be used by senior leaders, therefore it is needed to start data construction in terms of the thinking of the "big data", to integrate available data and meet the real-time needs of leadership decision-making with characteristics such as large amount, high speed, variety, authenticity, circulation, and interactivity.

To achieve value transformation, it is necessary to use many technical tools, such as BI, R language, etc. Without these technical tools, it is impossible to present the value of data to users and effectively support leaders' management decisions, while technical tools cannot play their proper value without the factor of data. Therefore, presenting the value of data and the construction of big data platforms necessarily encompasses the construction of big data processing and business intelligence application analysis. Big data is characterized by the dynamic processing of streaming data, the core of which is the continuity and rapidity of real-time dynamic data analysis. The final decision support system should be displayed through tables and icon graphics, as a data chart report with detailed classification, colorful marks, and authoritative data is the best way to present to the leaders as their decision support.

---

[4] TANG Tao. Research on the Analysis Method of Internet Public Opinion Based on Big Data [J]. Journal of Modern Information, 2014, 34(3):3–11.

4 Building a Scientific System for Data Decision-Making

What is data used for? Data is first and foremost used to make decisions. Human decisions may not always be rational, but we have deduced a lot of assumptions and judgments through data, and at least there are still many organizations that place more emphasis on rationality when making decisions. Therefore, building a scientific and rational data decision-making system can start with the following.

1. It is important to be aware of the value generated by big data and to do a good job of screening and evaluating big data assets.
2. View the design from top-level, and make good system planning. The change in leadership thinking is particularly important. The era of big data requires to be advancing with the times and speedier so leaders should pay close attention to the impact of big data on enterprises from an ideological perspective, treat data as a core resource for organizational development, and treat the collection, management, analysis, and effective use of data as a major task for building core competitiveness; top-level design and system planning should be carried out as early as possible; information technology should be given full play to the important role of information technology in supporting data analysis.
3. It is important to both strengthen data management and data security.
4. Optimize internal coordination models and strengthen external cooperation for win–win situations. Interconnected coordination and collaboration must become a management norm and must be the management model that leaders should follow.
5. In the decision-making process, the process is more important than the analysis of data. Many top analysts can produce insightful analysis, but the lack of an effective decision-making process can also lead to ruined results. Establishing digital processes in an organization is far more important than building a strong data system and data analysis team.
6. The attraction and cultivation of professional talents. To programmed decision-making process requires not only organizing decision-making in a systematic way, but also applying scientific methods, binding decision-making activities to established procedures, avoiding decision-making being influenced by the knowledge structure, mood swings, emotional impulses, and value preferences of decision-makers, and making any decision an optimal choice that drives development.[5]

Big data management has become an important issue that many government agencies, enterprises, and institutions must focus on. It has come the best time now to build a big data integrated management platform, which can help enterprises, institutions, and government departments effectively manage big data and make it a powerful tool for leaders in management and a driving force for organizations to move forward.

For think tanks, the opening up of data not only means that researchers can use more adequate resources to enhance the scientific and accurate research and

---

[5] JIAO Yiwei. Standardized Management for Enterprise [J]. Chinese Township Enterprises, 2008(6).

consulting results, but also cooperate with various data companies to discover features or patterns that were previously overwhelmed by the massive amount of information through big data analysis, so as to give different analytical judgments and countermeasure suggestions, or even different countermeasure suggestions at the same time, in order to better match the needs and personalities of different leaders, thus providing effective support for government decision-making needs.

5 Receive Data Results, then Make Solutions

With the help of big data technology and thinking that "everything can be quantified", government departments can obtain more information than before based on the huge amount of management and service targets, and use data mining technology and experience to achieve more accurate insights and predictions, thus greatly enriching the means and methods of government and social governance. It has now become a consensus in the field of management science that "there's no management without quantification".

Using methods such as data mining to mine vast amounts of data, computers can present us with a world full of correlations and then predict the likelihood of things happening through correlations. A correlation may not tell us exactly why something is happening, but it will alert us that it is happening. In many circumstances, such help as a reminder is valuable enough. For example, inferring the likelihood of congestion on a road from point data on the operation of public transport vehicles, finding the likelihood of tax evasion from unusual data characteristics of taxpayers, inferring the likelihood of a flu outbreak from the keywords people search online, etc.

With the aid of big data, the government can, on the one hand, comprehensively perceive and predict the various services and information required by the public in real-time, discover the hotspots in demand in time, and turn the superficial demand judgment into the perception of demand details, allowing the government service provision more accurate and personalized, and providing users with more intelligent and convenient services; on the other hand, the sign of the change of modern government public management is people's satisfaction and sense of security. By monitoring data and making timely multi-dimensional and multi-level subdivisions, the government can promptly and comprehensively sense how citizens feel about the government's work, thus enhancing the public's satisfaction, sense of accessibility, and sense of well-being.

In September 2013, the Public Transport Department of Beijing Municipality launched the "Custom Bus" platform, where people can put forward their travel needs and the bus group can design business shuttle routes based on demand and passenger flow, then provide services such as passengers recruitment, seat booking, and online payment on the platform. The business shuttles run according to the agreed time, place, and direction, guaranteeing seats for each passenger, while the daily cost is also far lower than driving or taking a taxi.

The HOME Social Risk Prevention and Control and Big Data Perception RAAS Service Platform, launched by China Statistical Information Service Center and Home Big Data in 2017, can scan the security perceptions of 31 provinces and cities, more than 400 municipal districts, and more than 3,000 county and urban areas across China, and dynamically grasp the predisposing factors affecting the public's security perceptions in each area. Local governments or relevant departments can query the level of local security perceptions, analyze the main factors affecting security perceptions and track the main hotspots affecting security perceptions on the platform, then further conduct in-depth learning and effective early warning on big data by combining data from local comprehensive governance work, grid-based management, and other relevant data sources. With the gradual increase in data volume, the predictive role of big data can be gradually released.

Data sharing and opening up mean not only that more information can be used, but also that big data techniques and analysis can be used to discover more features or patterns that can better support the government's decision-making needs.

Receive data results, then make solutions. Building a scientific system of data decision-making has become a possibility in the era of big data.

Leadership in the Era of Big Data

Qing Jiang

1 Leadership as a Function of Management

Leadership, which means leading and guiding, is not simply about "managing people". The main responsibility of a leader is to stimulate the potential of one's subordinates, to make each of them work to 100% or even higher of their potential. The relationship between leadership and management can be understood as "thought" and "action"—management is striving to climb up the ladder of success, to do things wisely and effectively; while leadership points out whether the ladder is against the right wall and determines whether the right thing is being done.[1]

Clarifying the difference between leadership and management will help to define a person's role so that leaders can focus more on what leaders should do and managers can do what managers should do. Warren Bennis, known as "the titan of leadership gurus", describes leaders and managers in general terms as the relationship between "original creation" and "copy imitation", which is characterized by the comparison in Table 1.

In the era of the new economy, where value comes more from people's knowledge, there is an increasing integration of management and leadership functions. People look at a manager not just as someone who organizes work, but also as someone who gives a clear and unambiguous goal. Good managers must therefore organize their staff, not only to maximize efficiency but also to take responsibility for developing skills, developing talent, and producing results. Internet-based communication unifies organizational activities and places new demands on leaders: on the one hand, it allows the management of the organization to be brought together as an organic

---

[1] ZHAO Hongjun. Rumination on Leadership Theory [J]. Journal of Beijing Administrative College, 2006(5): 50–52.

Q. Jiang (✉)
Zhongnan University of Economics and Law, Wuhan, Hubei, China
e-mail: 18712919290@163.com

© China Renmin University Press 2023
Q. Jiang (ed.), *Digital China: Big Data and Government Managerial Decision*,
https://doi.org/10.1007/978-981-19-9715-0_6

Table 1 The difference between a leader and a manager

Leaders	Make innovations	Original creation	Focus on development	Focus on people	Inspire trust	Long-term view	An overview of the whole picture	Question on "what" and "why"	Challenge the status quo	The opinionated general	Doing the right thing
Managers	Engage in management	Copy imitation	Focus on maintenance	Focus on system architecture	Rely on control	Short-term view	Focus on the profit	Question on "how" and "when"	Accept the status quo	The standard good soldier	Doing well the thing

whole; on the other hand, it allows managers to understand the external environment through information exchange. Information communication makes the organization an open platform system that interacts with the external environment. The leader or manager is at the center of this information network and is responsible for its smooth flow.

The role and functions of leaders in the era of big data are beginning to change. With big data, government leaders are able to better understand the real thoughts and needs of the public, thus improving their ability and level of service to the public; they are also able to intuitively find the problems that are most strongly reflected by the public and then solve them; furthermore, they can let the public clearly see the entire process of government operations, which is conducive to better monitoring of government administration by the public and creating a transparent government. To meet the advent of the era of big data, leaders need to keep up with the times, make good use of modern information processing technologies, and apply them skillfully to their work, so that data can become a vain to better serve the public. Business leaders can draw direct and clear conclusions and make direct judgment and decisions through related analysis of big data. As big data relies more on correlation analysis of data rather than causal analysis of business characteristics of the enterprise, the focus is more often on data sensitivity analysis. As a result, business leaders can use big data analysis to find out "what it is" and make the right decisions, even when they are completely new to the business. Big data also enables business leaders to change their decision-making process from "hindsight" to "foresight". In the era of big data, the traditional decision-making process as "hindsight" is hardly adapted to the change since factors such as raw materials, production equipment, customers, and markets are becoming increasingly fluid and complex.[2]

For leaders, big data is a management tool for governing, and the precise wisdom of big data analysis will replace intuitive judgment. Both business and government leaders must accept this reality and put it into practice as soon as possible in order to keep up with the trend of the times and become wise leaders. Whether you like it or not, whether you accept it or not, after several years of criticism, questioning, and discussion, big data has finally ushered in its time.

2 Information Flattening Becomes Management Tool for Leaders

Most traditional management theories revolve around the organizational characteristics of hierarchical structures. Henri Fayol, the "father of management theory", who proposed "The Fourteen Principles of Management", believed that superiors could not override their hierarchy to command and subordinates could not override their hierarchy to report. This is a classic in traditional theory. But the hierarchical

---

[2] ZHANG Caiming. Data-Driven Decision-Making for Managers [J]. Enterprise Management, 2013 (11):110–111.

form of organization and the matching classical management theories have met with huge challenges in the modern world. For example, according to Fayol's theory, the instructions from the top decision-makers at IBM had to be passed through 18 management levels to the most junior executives, which was not only extremely slow but also resulted in predictable distortions and errors in the transmission process.[3]

The advent of the Internet and the era of big data has created the conditions for leaders to implement flat management. In this regard, XU Jihua, FENG Qina, and CHEN Zhenru addressed in their book "Smart Government: The Advent of the Era of Big Data Governance" that "For leaders at the top of the pyramid, grassroots management is a "black box", only the specific personnel know the most real situation, where information is not well reported and instruction is not well delivered. The only way to keep the pyramid running is to rely on the extraordinary insight of the leaders, or on the inner integrity of each individual." The great attraction of flattening information is that it is fair for everyone involved. Moreover, with the spread of information visualization, data is no longer difficult to understand or for technical specialists to understand only, while data analysis is no longer complicated and boring, but is increasingly presented in all its beauty.

Management in the era of big data is networked, interactive, and flattening operating. The generation, processing, and utilization of data are the result of the joint participation of government, enterprises, social organizations, and individuals of the public. Transcending the concept of hierarchy across traditional management will open up the minds of managers, giving the grassroots and the top management the opportunity to interact directly, sharing information in a flat manner, so that leaders will know the most real information on the front line; management will achieve hierarchical responsibility, so that leaders will know where problems are coming from, making it easy to trace responsibility. The interaction between online and offline, and the flattening of management, have become important characteristics of the leaders of organizations in the era of big data.

Communication with others must be two-way, as not just a conversation with someone else, but a ballroom dance that requires two people to come together. In terms of internal management in enterprises, positive communication requires managers to put themselves in the shoes of their employees, communicate with them equally, listen properly to their intentions, and ensure that the receiving information is correct. The more senior a leader is, the more one should communicate directly with the staff, as one-way communication is just conveying orders downwards with only token feedback from subordinates, which is not only unhelpful in terms of supervision and management at the decision-making level but also bound to frustrate the motivation of staff and their sense of belonging over time. A wise leader is first and foremost a good listener, who is able to obtain information from his or her subordinates and reflect on them. Listening effectively and accurately will have a direct impact on the level of decision-making and management effectiveness, and therefore on the performance of the company.

---

[3] GUO Jianguang. Constructing Flat Management System to Achieve Integrated Operation in Bao Steel Corporation [J]. Shanghai Business, 2004(1):21–24.

The effectiveness of leaders' works today and in the future is based on the flexibility of both information technology and personalization. One of the main responsibilities of leaders is to sift through the clutter of information to find effective information that is important and needs to be processed immediately, as well as to distinguish it from irrelevant and obsolete information. In the article entitled "Integrating Social Networks into Business Strategy" published in Enterprise Management in 2013, it was argued that: at the management level, a business is an organization made up of people, and information is disseminated in and between the enterprises through people—the most fundamental activity in business operations. Although there are many popular social networking platforms, from Kaixin, Renren, and Weibo to WeChat, going up and down, the key elements of all these platforms are people and information, as the fundamental purpose of social networking is communication and sharing. Business social networking strategy is to address the issue of how to allow the people associated with the company access to the needed information securely at anytime, anywhere and in the shortest possible time, at the lowest possible cost.[4]

Minor or general traffic accidents on the roads in large cities such as Beijing no longer need to wait for traffic police and insurance company managers to deal with at the scene but can be determined by both parties to the accident by uploading photos themselves through social networks... Social networks are no longer just ordinary chatting and photo-sharing between people, but a platform for social networking between people, between organizations and people, and between organizations. The construction of social platforms has also made it easier to develop leadership. Leaders in the new era can share their ideas, practices, and various experiences through different information channels. As the world becomes flattened, the convenience and impact of information transmission are greater than ever before.

In the era of the Internet, as the distance between leaders and the public gets closer, respect for public opinion has become a sign of progress for the government.

3 Leaders Should Take the Lead in Improving Data Quality

In 2016, big data has fully penetrated all areas of society. Big data has become the key to improving the efficiency of organizations' operations and determining their development.

After the country introduced the big data development outline, various departments have taken positive action, introducing a variety of programs on big data, and the whole industry got passionate about big data. However, from my years of experience in the industry, these programs may look very good but still is a distance from the ground application. Why is it? The lack of specific implementation plans and specific executive references, the application level is difficult to see or enjoy the

---

[4] ZHANG Yanlei, LIU Lian. Integrating Social Networks into Business Strategy [J]. Enterprise Management, 2013 (9):104–105.

exact benefits brought by big data. In fact, we all know that no matter where the development is, it should be problem-oriented and application-oriented, which is the most basic common sense. Though in the early stages of big data development, it is understandable that being difficult to require experts or leaders involved in developing programs to have direct experience and front-line experience in the practice or even just experiences facing obstacles, or rather, those experts or leaders not yet have a certain data intelligence quotient (DQ)—data quality.

The point of big data is not about its huge volume. Many start-ups are now talking in terms of how massive the data we have is, but large scale without mining and application will just be a pile of rubbish, and the larger the scale is, the bigger the dump will be. Many places and government departments now have their own data centers, but when you actually think about it, not much data is really utilized, instead, they have to invest a lot of money every year to look after and maintain the centers, resulting in a huge waste. The data could have been mined for value and applied for benefits. In line with academician LI Guojie's view that "the development of big data should not be pursued on a large scale, but should be on the application first", my team and I have always researched and engaged in the practical application of big data, and positioned our work as "tools for managers".

I have always put myself as a practitioner, "preaching" on the cross-border integration of big data thinking, but cross-border integration requires a long process, and no one can be an exception. In the process of the full penetration of big data into all areas of society, I am very pleased to see that many experts and celebrities with considerable academic levels and popularity in other fields are giving lectures on big data one after another, reflecting the trend and inevitability of the full penetration of big data into all areas of social life.

With the hype of big data concept booming, some local government leaders actively rushed to the front line of the development before getting clear about the true connotation and value of big data, building exchange houses, data centers, holding conference activities, and turning the direction of investment to big data… Such preliminary work is also the necessary basis for the development of big data, as its agitation and appeal showed a positive impact on the full penetration of big data in all areas of society, but to really achieve application, just relying on publicity and propaganda is far from enough.

As a leading official, entrepreneur, or anyone who wants achievement in their career, it is impossible to succeed without intelligence, emotion, and courage. However, today when the full penetration of big data in all areas of society and social life, I believe that improving the data quality of leading officials has become the current need. In 2015, when I went to the Nantong Municipal Bureau of Statistics to give a lecture on "Big Data as a Decision Aid for Local Party and Committee Governments", I first proposed that leading officials should improve their data quality, which is also known as "Data Intelligence Quotient" that should be another important factor for a person to achieve success in their career. I believe that the data intelligence quotient is a comprehensive assessment of one's ability of knowledge, understanding, application, and degree of effectiveness on data. The data intelligence quotient is one of the main manifestations of a person's differentiated data decision-making ability and will play a decisive role in one's way to a higher level and making more important decisions.

In an organization, a person requires mainly intelligence quotient to do his own job, which consists of the personal ability of professional knowledge, and work experience; a person requires not only intelligence quotient to develop in an organization, but also the emotional quotient and the daring intelligence quotient, while the first determines one's leadership charisma of integration and coordination, and the second one shows his leadership power facing the challenge, competition, and risks; and if a person wants to make a difference in the career and in the organization, to make certain breakthroughs and exceed predecessors, the person must have a high "data intelligence quotient". The data intelligence quotient corresponds to the decision-making power of a leader, which allows decisions to be more scientific and objective. Today, as the government promotes the development of big data, leaders in all positions must take the lead in improving their own data quality—data intelligence quotient.

4 Super Brains and Talents Team

Can big data be a super brain and think tank for business and government leaders? The answer is yes. The popularity of mobile social networks has increased the interaction of information between people, and the emergence of social media has accelerated the speed and scope of information dissemination; the expression of public opinion and the transmission of information by the public and key opinion leaders can be quickly identified and analyzed by technology, and thus become the data basis for leaders' decisions.

In the past, because of the scarcity of data, major decisions for companies relied almost exclusively on the experience of managers themselves. In the era of big data, the information needed can be obtained from all levels, which means that managers must rely on their front-line staff to improve their decision-making; at the government level, as every one of the publics can enter the world of big data and become "data analysts", the main body of decision-making has shifted from the think tank elite to the public. Therefore, everyone in a big data environment is a data body, and the numerous data holders surrounding the leaders are the invisible think tanks. The participation of all people in social governance has become an important feature of leaders' decision-making in this new era.

Through the Data Cube service of Taobao, shop owners can understand on Taobao the macro trends of the industry, the sale situation in the market of their own shops, and the behavior of consumers in buying products, thus making rational decisions on the production of popular products and the digestion of stocks accordingly. At the same time, more consumers can also find their favorite products at a better price with the application of data. Intime has installed free Wi-Fi in its department stores and shopping centers, gradually capturing the data of in-store users and VIP users, and using the Intime website (www.linkheer.com) to connect offline physical shops and online VIP accounts. When a registered consumer enters a physical shop and his mobile phone connects to the Wi-Fi, the backstage can recognize it and all his past interactions and preferences with Intime will be presented backstage. Through the analysis of the e-tickets, walking routes, and staying areas of the consumers in the physical shops, the shopping preferences of the consumers can be identified, and some habits of shopping behavior, shopping frequency, and matching categories can be analyzed. In addition, the Intime website can even accumulate data on how different users like brands and discounts, relying on relevant data from mature shops and the user analysis of the city where the new shop is located, to export the guidance of stocking and investment attraction for new stores, and etc.[5]

In relation to our topic, the traditional management decision makers are the "elite executives" and "business experts", not the general public. With the growth of social media and the application of big data, the real influential decision makers are shifting from experts and elite executives to the general public who owns the data. Big data is rapidly evolving from an "optional" fringe to the "must-have" core. The battle for talent is on, but the jobs related to big data require interdisciplinary talent who can master aspects of knowledge such as management, statistics, mathematics, machine learning, data analysis, natural language processing, and even communication, public relations, etc. Face the reality that lacks talent in big data, and the needs for innovation in talent training mechanism, mobilizing the enthusiasm of local government departments are now the most urgent task. Firstly, a joint training mechanism for talents needs to be established. Talents in big data need to be jointly cultivated by the government, industry, and university institutions. The government, industry, and universities should jointly draw up the knowledge system in the field of big data and the areas of major applications, and jointly train the needed talents with the application as the goal. Secondly, innovative modes of cultivating talents need to be created, such as MOOC (massive open online courses) or competitions. Finally, local officials need to be mobilized. The value of big data lies in its cross-circulation. In addition to professionals, local officials should also receive training to improve their own big data quality and enhance their ability to utilize big data, so as to give full play to the role of big data in the governance of the country.

Do not expect one chief data officer (CDO) or one data scientist to nail down everything, but also try to put together a great team of data analysts, machine learning experts, and data engineers under their leadership, because the power of a team is always incredible.

---

[5] LEI Liang, PENG Zhen, LI Hong. Study on the Application of Big Data in Regional Brand Marketing [J]. Library and Information, 2015(2):77–81.

Enterprise Management in the Era of Big Data

Qing Jiang

1 Big Data as "Glue" of Team Cohesiveness

Tapping into the hidden social networks within an organization with big data can analyze how each individual in a team interacts with each other, through which managers can take more effective action to dramatically improve employee performance and business effectiveness. The power of big data analysis can be stronger and bring in unexpected and multiple changes to the management mode of enterprises. For example, through big data, management can understand how employees work and how team members work together in a project, and in turn, through the analysis of big data, employees can accurately pinpoint their role in the team and choose the most appropriate job position to make the most of their strengths and improve their productivity.

Today, social networks are transitioning to the workplace in a natural way by extension. Traditional communication methods have been out of date and the way leaders communicate and influence others has also changed. As a result, an increasing number of companies are transforming themselves in a common and distinctive way—socialized. By making use of the convenience and flatness of the Internet, companies can communicate with their internal staff with zero distance. Decision-makers in the organization can be the first to see all developments and ensure timely and correct decisions; employees can also discuss work topics with colleagues and managers anytime and anywhere without time and space constraints, which makes a truly barrier-free communication dialogue a reality. To learn to communicate, leaders must first actively build and participate in social networking, becoming a part of it to gain more trust.

Home Big Data has developed its own internal working flow management system in order to effectively manage the projects carried out. In addition to making full use of

---

Q. Jiang (✉)
Zhongnan University of Economics and Law, Wuhan, Hubei, China
e-mail: 18712919290@163.com

© China Renmin University Press 2023
Q. Jiang (ed.), *Digital China: Big Data and Government Managerial Decision*,
https://doi.org/10.1007/978-981-19-9715-0_7

DingTalk, WeChat, and email for regular interaction and communication, the working management system acts also like a scanner, storing and managing each work at any time in forms, with all employees operating on the platform. Throughout the day, there are visual data on the length of their work, the time spent on work content, and the efficiency of their work. Departmental managers, can make full use of these interactions and work data to personalize and match the work status of each employee and measure their working ability so that they can make timely and effective work adjustments and identify unsuitable employees and good employees. Although the use of the system is still being explored, it is already showing some results.

Social networks can be widely used in the future in the management of organizational operations: virtual meetings, work diary sharing, online issuing of tasks, leave approval, etc. Therefore, enterprise management should follow the pace of big data and fully use and utilize social sensors to enhance interactive communication, while effectively collecting and summarizing the real-time work dynamics of employees and conducting targeted analysis. At this moment, you will find that by clearing the barriers of internal communication within the team, team cohesion can be well enhanced thus business efficiency can be improved. Big data has become the "glue" that helps enterprises optimize customer service, expand their research and development, and stimulate team creativity.

2 Big Data as "Steering Wheel" of Human Resource

Big data will trigger a new revolution in the field of human resource management and drive it to a new level.

Human resource allocation means putting the right people in the right places. This is not something easily done well. Knowing exactly what a person is suitable to do requires an assessment of their abilities and qualities, as well as a specific and áccurate description of each job position. To this end, two aspects need to be done: firstly, a scientific assessment of employee quality and competencies; secondly, a meticulous description of each job position. One of the tasks of the human resources department is to strive for a high match between these two aspects.

Organizations in the community vary, as some are smaller, and some are larger. Some need top-level talents, others need just ordinary talents. For organizations with needs for top talent, a recruiting search for top talents should be carried out. In order to be professional, such activities are usually carried out through headhunters. The reason why foreign headhunters are able to find top talents is that they are backed by big data, which is a huge data warehouse supported by a very fast-calculating search engine that can find the right talent as quickly as possible. Using big data methods, headhunters are able to build their own "talent radar". Talent radars are based on a wide range of data sources, such as repositories of academic papers, records of speeches at professional forums, the number of papers in different academic fields and citation indices, important leads on social networking sites, etc. The headhunter's radar targets these objects from scans and searches for useful information that can portray the needed talents. Thomson Reuters uses the vast amount of valid data it collects to constantly report and recommend outstanding talents across a wide range of industries and sectors. If someone wants to know more about one exact talent, they have to pay for it. This is no doubt a great way for organizations that are eager to find the right people.

Big data can effectively help companies optimize customer service, stimulate creativity, and expand research and development. From now on, companies that do not know how to utilize big data to form efficient teams will be gradually eliminated. Both management and employees in the companies should pay attention to the power of data, and the changes of the times brought about by data, and strive to develop their own big data mindset. A large number of data traces are generated within companies. By analyzing the communication data between employees, it is possible to understand not only the individual performance of each employee, but also the cooperation status of the team, so that effective measures can be taken to improve the internal efficiency of the team, and even possible to select the right people to participate, and also to predict the cooperating situation between members and possible problems before a team is formed.

The appraisal is the core of human resource management. Without appraisals, an organization cannot achieve its goals. However, it is not easy to get the appraisal right. Human resource appraisal in the era of big data requires everyone who attends the appraisal to keep a personal work log, which means the person has to record what he/she does and how well he/she does it every day. In this case, managers can always know through the management panel what and how the person has done, and whether they have met the requirements and standards. Assistance can also be readily available if they encounter any obstacles to their work. This changes the previous problem of not having a timely overview of the situation and not being able to remedy it afterward.

For the circumstance of some e-commerce companies, big data can also predict the performance of each employee in advance. For example, whether or not a merchandise sales target will be completed, which used to be measured at the end of the year, or so-called "settle accounts after the autumn harvest", can now be predicted in advance, and staff can be guided in due course. So how do managers know which people are likely to fail to meet their targets? It turns out that they build a model using big data methods, through which link three figures of data: the first one is the "inquiry" price, which is the price of the item for click-on to inquiry about; the second one is the "item price" at which the order was placed for purchase, and the other is the "transaction price" at which it actually took place. There is a certain proportional relationship between these three figures.

The salary is a matter of concern to every employee, and negotiations about it are becoming increasingly common. It comes to an agreement if you would like to accept the standards given by the enterprise, or you can leave if you refuse. In fact, this is a kind of gaming process between employers and employees. In this process, how can we tell the level of approval of each other? This is important. Alex Pentland has developed a "sociometer" that records the signals that people output and process unconsciously, which are predictable. In a salary negotiation, if the other person is attentive and mimics your behavior, this means that he will play the role of a "collaborator"; if the other person speaks first and show reluctance to adapt the way he speaks and a tendency to slow down, this means that he will play the role of a "leader"; if the person is ignoring the conversation, moves his hands a lot when he speaks and appears fidgety, this means that he plays the role of a "discoverer" and might accept some suggestions; but if the person shows an attitude of "listening actively", the meaning is obvious. Only 30 s of social relationship measurement data can predict what role each person will play in the communication. This big data model is up to 95% accurate.

Talent loss occurs in both the enterprise and government sectors. In the past, we heard a lot of the phrase "retention by welfare, affection, and career". However, no matter how hard an organization tries, there is always talent loss. To reduce this, the organization can rely on big data to capture the overall dynamics of the workforce and the outstanding talent. As the saying goes, "when the moon is waning then the wind is blowing, the foundation stone is moistening then the rain is falling", there are signs and symptoms of everything. The human resource department has to carry out a timely and comprehensive analysis and dynamic observation of the working conditions of people in the organization. For example, the staff who had been working normally have been absent on leave frequently recently; the staff who were usually in good health have often been saying that they are not feeling well or even claim to be sick recently; the backbone who had been actively suggesting ideas are now silent. All these phenomena should be detected through careful observation. The human resource department should have the data sensitivity of a "data analyst" and be able to detect the early signs and take appropriate measures.

3 Big Data as "Supportive Partner" of Financial Management

The value of the application of big data by enterprises is particularly evident in the area of financial management.

Cloud computing and big data have become effective tools for financial management, making complex things simpler and thus helping finance to make decisions and manage more effectively. The "financial officials" in the enterprise might count only a number, but the number represents a huge amount; they might have never been to the location of the project or participated in the discussion of a project, but keep the most important gate at the beginning of each project; they are not in the front line of business, but know the profound reasons behind the problem; they have never directly involved in the market activities, but are of increasingly important as value creators for the enterprise. Now, they are becoming the pioneers of innovation in financial management.

Enterprise Management in the Era of Big Data

In a paper published in the journal Market Modernizing, it is addressed from the perspective of enterprise financial management that big data provides an opportunity for finance to transform from simply "bean counters" to management accountants. Previously, finance provided managers with a basis for decision-making by analyzing data. However, data analysis based on financial statements could only provide managers with extremely limited information. In the era of big data, companies are facing an increasingly wide range of data and a more complete casual chain between data. Through the seemingly ordinary data, financial managers can gain a more comprehensive understanding of the current situation and problems of the enterprise in the process of data analysis, and evaluate the financial situation and business results of the enterprise in a timelier manner, thus revealing the contradictions and problems in business activities and providing directions and clues for improving business management.

With the help of big data technology, financial managers can also effectively improve financial management, reduce costs, and bring in substantial profits for the enterprise. In this way, big data offers distinct opportunities for finance professionals to create value. For example, by utilizing big data technology to penetrate and analyze the product sales revenue in the projected income statement, financial managers can obtain detailed data information for different periods, different product categories, and different classification criteria. By comparing the actual data and projected data of the enterprise and designing a set of optimal management solutions for the future operation of the enterprise on this basis, the "best allocation" of resources and the "maximum return" on future performance can be achieved. As a result, big data expands the horizons of finance to important management areas such as decision analysis and support, risk management, credit management, and operational cost management.[1]

The ERP management system is the operational model of modern business management. It is a company-wide system that is highly integrated and covers the management of clients, projects, inventory, procurement, and supply, etc., maximizing efficiency by optimizing the resources. According to CHEN Dengkun, former Vice President of Kingdee International Software Group, ERP management systems were only used by professional users in the past, but now they are used by all staff, even the CEO. It was pursuing the efficiency of the business in the past, but now it is the efficiency of all staff; it was about processes and rules in the past, but now it is about sharing and assistance; it was used solely in the office in the past, but now it can be used anytime and anywhere. With the help of Internet technology, it is also possible to obtain timely and accurate corporate information, thus providing assistance to the company's decision-making. For example, order information, payment information, and account information can all be automatically collected and generated and can be compared intuitively according to different indicators, so that the reliability and relevance of the data are significantly improved. In turn, the utilization of big data analysis, helps top management to gain insight into big trends and grasp the future.

---

[1] ZHANG Xiaolei, FAN Xiaoming. An Introduction to Financial Management in the Era of Big Data [J]. Market Modernization, 2014(2): 166.

According to JIANG Chao, the Executive Director and President of China Wireless Technology Limited, "Firstly, big data can quickly help the finance to build financial analysis tools, rather than just doing the accounts; secondly, big data should not be limited to only a company's own big data, but more importantly, learn how to use other companies' data. For example, a small company may have no idea about classifying wages and salaries, as well as the salary level of departments. Later on, through getting a salary analysis done by a consultancy, a reasonable position was given. The reason for this is that consultancies who have database storage and a huge source of data can help enterprises manage their operations better."

Analyzed from the perspective of government accounting management, big data technology offers a wide scope for accounting managers.

Guangdong Province has consolidated the information of the province's 1.8 million accounting licensees into its accounting management information system, expanding the breadth of data collection, promoting the depth of data mining, and increasing the use of data. Collecting, storing, and analyzing data are the three necessary segments to discover and extract the potential value of data. Relying on the current accounting management information system, the Accounting Department of Guangdong Province has built a credit system for accounting practitioners, which records the credit status of every practitioner in detail. All the misdeeds of those accountants who are involved in false accounting and illegal practices are recorded in the credit file and publicly exposed, which is of great significance to effectively promote accounting management. At the same time, the tutorial materials required for the Accounting Qualification Certificate Examination and the relevant notes on auditing standards for Certificate Practicing Accountants during the annual audit can be accessed through the accounting management information system. With the help of a big data platform, Guangdong Province provides accountants with a large database that brings together accounting information.[2]

Dongfeng County in Jilin Province used big data technology to build a "centralized financial management system for administrations", realizing a new financial management model characterized by "full monitoring of financial expenditures, centralized accounting data, and automatic generation of departmental accounts". Big data technology has made the country's financial system data more realistic and the accounting and supervision model more standardized.

4 Big Data Powers Marketing Change

In the era of big data, companies are already facing a very serious challenge. With the coming of big data, many brands might be overturned and replaced, while many traditional brands may not be seen in the next few years if they don't keep up with

---

[2] QU Tao. Accounting Management in Guangdong Launches "Big Data Strategy" [N]. China Accounting News, 2013-04-26.

Enterprise Management in the Era of Big Data

the times or "cannot recognize the ongoing situation". In fact, it can be seen the happening of such a phenomenon, which just has not yet spread on a large scale.

In terms of conventional understanding, a brand is a sign of recognition, a symbol of spirit, a value concept, and a core expression of its product's superior quality. A brand is an abstract expression of communication between an enterprise or a product owner and its consumers through various marketing methods. The brand value originates from the consumer, which seeks general awareness among the consumer and then consumption, rather than just focusing on marketing and promotion. As technology further drives the development of the Internet, big data presents new challenges for brands.

In the face of the various changes coming with big data, both enterprises and consumers should keep up with the times. In the old days, enterprises had absolute control over information and could collect data in a variety of ways. They conducted research and summarized results for further use, thus affecting consumers, while consumers could only accept. In the 1990s, Wahaha and Hengyuanxiang were brands that grew rapidly through television advertising, and at that time, enterprises took the initiative and consumers could only listen to what they were told. Therefore, back 20 years ago, enterprises had the initiative to deliver information and consumers had no choice even though it might not be the information they would like to accept.

What about the current situation? This is the age of the consumer, the age of the real customer first. The Pad, PC, cell phone, wearable device, set-top box at home, and other electronic tools that everyone has are the same channels as owned by enterprises, and there is no difference in the channels and amount of information obtained between enterprises and consumers. The more you know, the more transparent the information is. In other words, all consumers have become smarter. Nowadays, consumers don't just listen to whatever enterprises say, but will definitely think about it. Therefore, consumers now have the basic conditions to compete with enterprises in terms of access to information, and under these new conditions, it is easy for enterprises to develop brands as long as they meet the needs of consumers. Xiaomi and Taobao, as we all know, are brands that have been successful because they are truly consumer-focused and can fully satisfy consumers' needs, they have developed their own brands by using Internet interaction rather than advertising. The environment for both consumers and businesses has changed, and the genes of change are brought about by the Internet. Enterprises need to learn to be consumer-focused, to shift the focus of the market from products to users, from supply control by businesses to active response to user needs, from convincing customer purchase to allowing users to deepen their understanding of products, and from brand communication to brand dialogue.

5 From Brand Communication to Brand Dialogue

The value of big data for brands does not lie in the data itself, but in the way of thinking to dig deeper into the value behind the data, to solve specific problems, and thus realize brand communication towards brand dialogue.

5.1   Learn to Use Big Data Thinking

Big data is used to solve problems. Everyone is currently being collected by various organizations for data such as operator information, bank information, hospital cases, etc. Is this data being collected and fully utilized? Seems not. For example, every time I go to the hospital, I still have to start all over again, so why not implement the sharing of an electronic casebook? Why do I have to buy a new casebook every time I go to a hospital? Therefore, it is still not that many using big data to solve problems, and even some organizations that own large volumes of data are not using them to solve problems.

This is actually a mindset at work. For brand management in the era of big data, business leaders also need to use big data thinking to solve problems. Many government departments, corporate organizations, and even individuals are now conducting training related to big data to develop big data thinking, discarding the previous traditional anachronistic thinking habits and shifting from serial thinking of causality to correlative parallel thinking. Today's big data has made relevant research more valuable and has transformed the way of communication and paths for brands, so the first thing that big data should change is the mindset of brand managers.

5.2   The Value Behind Big Data

Big data is not a concept, but rather an effective means of solving problems. If there is only data without developing applications and insightful analysis, it is just a pile of data junk, and the bigger it gets, the more troublesome it becomes. Some discussion on big data in China now mainly focuses on two ends: one is on the cloud, talking about concepts and inspirations to make you passionate; one end is on the server, talking about Hadoop, Java, and C Language that sounds obscure. So, have these two ends only existed after the concept of big data? Apparently not, they have always been there. Can these two ends solve problems if they remain unrelated? Of course not. Only when solid implementation was made for concepts on the cloud, and technology within the server was applied to the practice to solve the problems that need to be solved, is the complete business form that we are committed to.

Now with big data, the way to build a brand has to be different from traditional. Google and Baidu have a natural advantage in terms of big data in that they have databases storing demand-based search data. If you want to know something, you can just Baidu it and the relevant information will come up, and then analyze and determine. This is very valuable for understanding demand and predicting trends.

5.3 Big Data Makes Brand Management More Real-Time and Stylish

Big data has made brand management more real-time, more stylish, and more personalized. When lots of enterprises conduct new media promotions these days, their creativity is real-time and dynamic. This is because the message needs to be sent out in a timely manner now, otherwise the consumer's focus may have shifted after one night. Especially in an environment where social marketing has become the mainstream, it is essential for brands to build good customer relationships if they want to promote their brand awareness and popularity in the market, and the role of big data has become increasingly important.

5.4 Big Data Analysis Can Make Brand Marketing More Effective

For enterprises, understanding the needs and preferences of users can effectively help them target their markets and develop targeted brand marketing strategies. The influx of numerous data helps brands to determine the direction of product improvement and provide specific and relevant services to users. When a user browses the web or logs into a social platform, his every action is faithfully recorded by the Internet, including the time the user spends online, the browsing track, the content viewed, forwarded, liked, etc.; and as the timeline increases, the data of the user becomes richer, the behavioral characteristics of the user become an increasing number of obvious, and the quality of the user's data becomes higher. With the development and growth of modern digital technology, enterprises' analysis of user preference analysis has also become more accurate. There is no doubt that big data has made enterprises more aware of their users.

Brand marketers, have to face the same shift to big data as well. Before big data analysis, marketers had no way to track consumers' shopping patterns or to know what brands, products, and services consumers preferred. Big data allows him to gain new marketing ideas to make brands more responsive to market needs; enterprises can adopt different marketing strategies to provide a better product experience for their consumers. Big data analysis allows marketers to see market trends and correct marketing strategies and marketing directions for weak areas.

The marketer of the future will need to collect more data, extract valuable customer information through analysis, and improve the interaction between brands and users to make them more relevant to their consumers. It is inevitable that marketers who do not shift in time may be eliminated. If a marketer is still going to shops and trying to convince customers to buy products as before, that marketer is already outdated. Therefore, marketers have an important role to play in brand management in the face of the challenges of big data.

The widespread application of mobile smart terminals has made large-scale social communication a reality and has led to changes in the strategy, content, and form of traditional brand public communication. In the era of big data, the real value of big data is to find the entry point for brand public relations strategies in the mass of data and to communicate with consumers.

On 28 April 2017, the 4th Annual China Brand Reputation Conference was held in Beijing with the theme of "In the name of the people, give praise to quality products", and the conference released the fourth "China Brand Reputation Yearbook", and "Data Reference of China Brand Reputation in 2016" at the same time. The Yearbook covers three major categories of brands FMCG, durable consumption goods, and necessities and services, including 15 categories and nearly 60 sub-categories, such as household appliances, consumer electronics, daily chemicals, food and beverages, automobiles, pharmaceuticals, pregnancy, and baby supplies, and other industries that are closely related to people's lives, among which brands such as Sanyuan, AUCMA, Lelch, Philips Avent, International Children's Club (ICC) are listed as "Chinese Good Reputation Brand in 2016". The conference was attended by more than 300 experts and industry authorities in the fields of big data, reputation marketing, and public relations, and the speeches made by the participants further confirmed the inevitable trend of "moving from brand communication to brand dialogue".

WANG Haifeng, Deputy Director of the China Statistical Information Service Center, said that conducting research on and measuring consumer satisfaction and brand reputation as well as releasing a brand reputation index to society has become a regular work of economic activities in developed countries. In 1989, Sweden takes the lead in establishing and releasing the world's first national customer satisfaction index, then Germany, the United States, New Zealand, Canada, and South Korea established their own monitoring systems. The voice of "God" must be listened to carefully and the needs of consumers must be responded to actively. Market competition in the past was mainly based on quantitative expansion and price competition, but now it is gradually shifting to quality-based and differentiated competition, where enterprises need to pay particular attention to changes in the market, to changes in customer needs, and to brand reputation. The brand reputation shows the direct feelings of consumers towards the consumption of products and reflects the brand pursuit of "the customer is God", which is highly consistent with the development of the industry. The release of the results can even have an impact on the stock market and can be regarded as a barometer of industrial development. In the context of personalized and diversified consumption becoming mainstream, and in the new era of widespread Internet access, many consumers will judge by their own inner feelings and the experience of others, fully advocate in the quality of products and services, becoming a new element of brand management.[3]

---

[3] The 4th Annual China Brand Reputation Conference was Held, 2016 Brand Reputation Yearbook and Consumer Guide was Released [EB/OL]. China News. 2017-04-28.

Research on Brand Reputation Based on Big Data Technology

In 2013, the project China Brand Reputation Index (CBRI), launched by China Statistical Information Service Center (CSISC), is a big data interaction process based on user fragmentation behavior research by the team I led, and is also a demonstration of CSISC's exploratory achievements in the field of the application of big data research. The research results on reputation covering nearly 40 categories in the fields of FMCG, durable consumption goods, and necessities and service, bringing new inspiration and responsive development opportunities to the industry, society, and brand owners.

Reputation index evaluation system is a method of evaluating an organization or brand using multiple indicators in multiple ways, of which the basic idea is to follow the principles of scientific, orientation, comprehensiveness, comparability, and maneuverability, as well as selecting multiple indicators and conducting a comprehensive evaluation according to the different weights of each indicator. The monitoring vehicles are online media news, Weibo, BBS, blogs, communities and so on. The data acquisition method is: based on big data monitoring system developed by the China Statistical Information Service Center and intelligent web crawler technology, the Internet content monitoring system with domain as the unit and keyword as the direction, 24-h real-time monitoring of crawled information to get the information that meets the screening through manual sorting, and Baidu directed search. Some of the data used is stocked while the other is sample.

The Brand Reputation Index consists of six main indicators: brand awareness, consumer interaction, quality recognition, corporate reputation, product satisfaction, and brand health.

Brand awareness refers to the breadth of a brand or organization's information spread across the Internet, and is mainly based on data analysis from the China Statistical Information Service Center's Big Data Monitoring Platform and statistical analysis of search engine results. Generally speaking, the higher the brand awareness index, the higher the reputation index.

Consumer interaction is the extent to which a brand or organization is discussed in major online BBS and blogs. It is an indicator of how deeply a brand or organization is communicated on the Internet and is based on search engine results from a sample of websites. Generally speaking, the higher the consumer interaction index, the higher the reputation index.

Quality recognition refers to consumers' continued interest in a brand's product features, product quality and other contents after they have paid attention to the industry and brand. Generally speaking, the higher the quality recognition index, the higher the reputation index.

Corporate reputation refers to the degree to which a specific positive public sentiment event has improved and enhanced the brand or organization's online reputation; the higher the score, the better the organization's online reputation. The level of corporate social responsibility is also one of the key elements of reputation research. Consumers' choice of an enterprise's products is very often also depending on the enterprise's social image, which is intuitively reflected in the image of corporate social responsibility.

Product satisfaction directly affects the level of a certain brand's reputation index, the higher the satisfaction index, the better the reputation. The study takes product rating information data from major e-commerce platforms as the research object, and compares the total number of positive ratings of a product with the total number of all ratings in an index to calculate the product satisfaction index. In order to reflect the rating status of all brands in a comprehensive longitudinal manner, this index takes a historical base accumulation into consideration.

Brand health refers to the number of consumer complaints about a brand's products or services after they have been sold, which are reflected in various channels such as after-sales phone calls, online complaints, 12,315 consumer complaints poor e-commerce ratings, and etc. Various complaints erode the health of a brand like a lesion. A large number of consumer complaint information will visually reflect the health of a brand; the fewer the complaints, the higher the brand health. This index study mainly collects data of online complaints from the research subjects. The complaint index is the ratio of the number of consumer complaints about a product to the total number of complaints about similar products or services, which is regarded as the brand health.

The following is the Report on Domestic Infant Formula Brand Reputation Research of the first half of 2015 released by China Statistical Information Service Center together with China Quality News. In this report, it can be seen that based on the data compilation of big data, different companies show different core competencies and big data really makes brand marketing reach new heights.

Sanyuan Continues its Reputation Glories, while Feihe and Guanshan Rank at the Bottom of Blacklist
- Releasing of Report on Domestic Infant Formula Brand Reputation Research in the first half of 2015

Big Data Research Laboratory of the China Statistical Information Service Center (Home Big Data), together with China Quality News and Qianlong (www.qianlong.com), released the "Report on Domestic Infant Formula Brand Reputation Research in the first half of 2015" (hereinafter referred to as the "Report") on 17 July. The Report takes monitoring research on the 267 brands on sale on the Internet with more active market performance, among which 21 domestic infant milk power brands with active performance in the first half of 2015 were selected based on sound level for research, namely Edison, Ausnutria, Beingmate, Feihe, Guanshan, Guangming, Biostime, DUTCOW, Red Star, GARDEN, Junlebao, WISSUN, Nanshan, Arla Foods & Mengniu, Sanyuan, Synutra, Scient, Wondersun, Weichuan, Yashily, Yili (the above brands are listed in alphabetical order, as shown in Fig. 1).

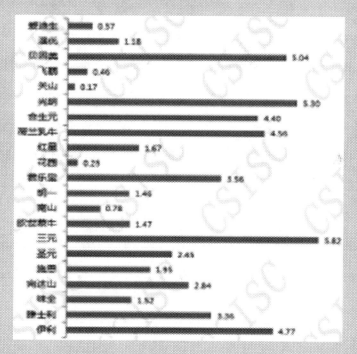

Fig. 1 Domestic infant formula brand online reputation index in the first half of 2015. *Data source* Big Data Research Laboratory of CSISC

From the overall online reputation index of domestic infant formula brands in the first half of 2015, Sanyuan, Guangming, and Beingmate lead with their different strengths. In addition, DUTCOW stood out like a dark horse in the first half year in its performance on product satisfaction and brand health index.

Compared to the previous year, Sanyuan continued to uphold its philosophy of "good quality" and achieved excellent results in terms of quality recognition and corporate reputation, winning the top spot again in the first half of the year, while Feihe which performed well in the previous year, saw its performance plummet due to a series of negative news such as product sampling failures in the first half of 2015. In addition, the activity and reputation performance of some small- and medium-sized brands in the field of formula, such as Guanshan, Herds Dairy, and Yabais, were also worse than before.

According to JIANG Qing, MBA co-adviser at Zhongnan University of Economics and Law and director of Big Data Research Laboratory of the China Statistical Information Service Center, competition in the infant formula industry today is fierce, with price wars and channel wars raging, so how can formula companies win back the consumers? Fancy marketing techniques or price wars are the choice of many formula companies, but in the opinion of many industry insiders, this is not a long-term solution. The image of a brand is overwhelmed by negative news of products; a low-price strategy may be able to gain attention and boost sales for a company only in the short term, but in the long term it will have very limited effect. The key to these problems lies in the fact that in the era of the Internet, the information asymmetry has been greatly eliminated, and consumers are becoming an increasing number of rational. The inflection points of the formula industry from price competition to quality competition has arrived, for formula companies, they will no longer win by low pricing, but ushering in the era of winning by product and reputation. The only way to shape a high-end formula company is to strictly control the quality of the product. The release of the report can, to a certain extent, make formula companies realize that taking the road of quality strategy is the effective way for them to develop in a long term. However, the introduction of these data must be supported by big data!

Big Data Promotes the Construction of Digital China

Qing Jiang

1 Smooth the "Last Mile" of Smart City

In May 2015, International Conference on ICT for Education and Training was opened and President XI Jinping said in his congratulatory letter that in today's world, technologies are developing rapidly with each passing day when modern information technologies such as the Internet, cloud computing, and big data are profoundly changing the way humans think, produce, live, and learn, profoundly demonstrating the prospects for world development.

In the 1990s, the concept of "smart city" emerged all over the world. In many developed countries, information on public services such as transport, water, electricity, gas, and gasoline are organically connected via the Internet to better serve the needs of citizens in terms of living, working, learning, and healthcare, as well as to improve government control of the environment and traffic management.

The purpose of building the smart city is to improve the quality of living environment, and big data is the "master" of it. In 2009, the Municipal Government of Dubuque partnered with IBM to create the first smart city in the United States. They used Internet of Things (IOT) technology to connect various city utilities in a neighborhood of 60,000 residents to respond intelligently and better serve citizens through monitoring integrating and analyzing data. The first step of Dubuque was the installation of CNC water and electric meters with low flow sensor technology to households and stores to prevent waste form water and electricity leakage. A comprehensive monitoring platform has also been set up to analyze and display the data in a timely manner, so that the entire city has a clear picture of how resources are being used through the data.

Smart cities in Europe play the role of information and communications technology (ICT) in urban ecology, transportation, healthcare, smart buildings, and other

---

Q. Jiang (✉)
Zhongnan University of Economics and Law, Wuhan, Hubei, China
e-mail: 18712919290@163.com

© China Renmin University Press 2023
Q. Jiang (ed.), *Digital China: Big Data and Government Managerial Decision*,
https://doi.org/10.1007/978-981-19-9715-0_8

livelihood areas, developing low-carbon housing, smart transportation, and smart grid, as well as improving energy efficiency, addressing climate change, and thus promoting low-carbon, green, and sustainable urban development. Singapore, on the other hand, provides real-time and effective traffic information guidance for citizens' travel through various sensor data, business operation information, and rich user interaction experience data.

In China, when eight ministries and departments, including the National Development and Reform Commission, jointly issued the Guideline on Promoting the Healthy Development of Smart City in 2014, the top-level coordination period of smart city construction was officially initiated. In February 2015, the Inter-Ministerial Coordination Group for Promoting the Healthy Development of Smart City was set up by 26 departments and organizations, officially starting the coordinated implementation of smart city work across the country. In recent years, the information barriers between departments have gradually broken, and some local application models in smart city have emerged. In Yinchuan, for example, with government information technology connected to 23 systems in 12 units, citizens can complete all applications at once through any of the service windows, while information is transacted between different departments, reducing the time it takes for businesses to apply for commercial registration from one month to one day. At present, Yinchuan has opened 515 Weibo accounts on party and government affairs at all levels, forming a large-scale and systematic operation mechanism, and building a three-tiered organizational system on Weibo for city, county (district), and town, as well as the public service system for water, electricity, heating, gas, and public transport that are related to people's livelihood. Whenever you encounter a problem, you can simply @WenZhengYinChuan and your request will be answered and solved in a timely manner, with a completion rate of 97.1%. At the same time, the government Weibo also sends private messages to netizens to rate their satisfaction with the administration and to monitor accountability for lousy work.

Then there are smart city management in Rugao, Jiangsu and Tongzhou, Beijing, relying on the "five-in-one" city management IoT platform construction of perception, analysis, service, command, and monitoring, relying on the city's high-definition cameras and grid-based teams to monitor and collect real-time associated data of city management that assist in better strengthening the service capacity building, as well as enhancing the interactive ability and emergency management to quickly respond to the demands of the public. It realized the leap from digital city management to intelligent city management, which improved the public image of city management.

However, we should also see that although China's smart city construction has been underway for nearly a decade, there are still no mature smart city cases. I believe that the core of a smart city is the application of big data, which can smooth the "last mile" of smart city construction. More importantly, the key to building a smart city is to achieve cross-sectoral and socially oriented openness of big data, and big data applications for various fields such as people's livelihood, in order to create a more livable urban environment, a smarter urban life, and a happier citizens' experience, so that to force the transformation of government functions and the reconstruction of administrative processes, which is the feasible path.

The construction of digital China requires the gradual completion of the smart city, which cannot be accomplished overnight. It requires local governments to understand top-level planning and develop plans based on the history, culture, resources, and strategic positioning of their cities. The government and relevant enterprises need to share the responsibility to build digital China based on the smart city.

2 Transformation of Traditional Government to E-Government

Since the 1980s, a new public management movement has emerged in Western countries such as the UK and the US with the aim of reshaping government departments. The advent of the era of big data has accelerated the change in public management in Western countries.

With the rapid development of Internet technology and data analysis technology, all elements of government, market, and society are showing new characteristics that are different from those in the past. For Western countries, the accelerated transition from traditional government to e-government is actually about improving the efficiency of government and making the limited government resources as much as possible to get the proper government management utility.

E-government is based on informatization, which means that the more informative a government is, the more developed its e-government will be. According to former US President Barack Obama, "The more the people know, the more accountable government officials are likely to be." The US has been at the forefront of open data and sharing in the world, with Obama proposing to push government data onto the Internet in a common, standardized format, "so that citizens can track and query information about government funding, contracts, earmarks, and lobbyists".

In May 2012, the US unveiled its Digital Government Strategy. Developments in information technology have enabled the goal of open data to be realized, allowing the public to access government information and public service information from any device, anytime, and anywhere. For example, the US established a shared database on taxation in 2014, through which taxpayers can access their personal tax records for the past three years to more easily take out mortgages and loans.

Open data brings open government, and the rapid shift from traditional government to e-government facilitates a more open and transparent public sector. Canada is also a leading exponent of the transition from traditional government to e-government. The "Government On-Line (GOL)" aims to provide online services to the public by consolidating various scattered information resources and organizing various public services in an aggregated manner so that the public can access public service information more easily.

The government's public services cover all areas of people's livelihood. Big data can help the government optimize public service processes, simplify public service steps, improve public service quality, develop the country's economy, and make people's lives happier. Government disclosure based on massive amounts of data safeguards citizens' right to know and provides the public with more comprehensive data services. Putting the power of information into the hands of the public, e-government in the era of big data has no doubt contributed to the credibility of the government.

Since the New Public Management (NPM) movement, the public sector in developed Western countries has focused on citizen satisfaction, pursuing efficiency, and implementing clear performance objectives. By adhering to a service-oriented approach, governments have moved away from being superior dictators but rather become providers of quality and efficient public goods. Advances in information technology and data analysis have created the technological basis for better and more efficient public services, enabling governments to use more advanced technological tools to change the way they manage public services and achieve previously difficult public service goals.

Emergency management is an important element. In the era of big data, using databases and sensors on a large spatial scale governments can quickly access data on geography, population, and disasters, so that more quickly lay the foundation for disaster response and relief. The United States has installed hundreds of observation instruments at the Yellowstone Caldera. The data observed by the instruments is divided into regular data and anomalous data, and the more anomalous data there is, the greater the likelihood of a natural disaster occurring. The observational data is transmitted in real-time to the early warning system, which rapidly screens and releases data to the public via the Internet. The tsunami warning after the "3/11-The Tsunami" in Japan is also an example of the utilization of big data. After the earthquake, the US National Oceanic and Atmospheric Administration (NOAA) quickly issued a tsunami warning. The reason for this rapid response lies in the fact that the US has established a large network of ocean sensors with global coverage. Through these sensors, the NOAA was able to access and analyze a large amount of ocean information in a timely manner, facilitating the timely issuance of disaster warnings and gaining time for the public's personal safety and property relocation.[1]

In addition, as the cost of data storage decreases and the speed of data reading increases, police are able to store more social information for law enforcement and crime alert purposes. For example, the Los Angeles Police Department has applied "Crime Mapping" based on data analysis to its daily work, strengthening patrols in areas with high occurrences of criminal behavior. This has effectively reduced the number of crimes in the district and ensured the maintenance of law and order and safety in the district.

With the development of the Internet, different organizations and sectors are becoming an increasing number of connections and the interdependence between the country and society is becoming stronger, which brings a very good basis for social governance, national innovation, and economic development, which is based on timely and even real-time communication.

---

[1] MA Li. Changes in Western Public Management in the Era of Big Data [N]. Study Times, 2014-12-08.

Western countries have developed a number of apps that can be used by the public, and the data from their use is then collected and analyzed backstage. For example, the US Health Service has developed apps for epidemic diseases, which are tied to people's daily lives through medical consultation and treatment. With such a foundation of information and data support, the government will be able to manage society with great ease. At least the government will be able to know at the first instance what is happening in society, what the causes are, how to respond, and what the future might be.

According to Dutch scholar Walter, "The main challenge encountered by public management as governance is dealing with netted environments, which is the interdependent environments." And the era of big data offers further possibilities for dealing with such netted environments.

Firstly, there was a deepening of collaboration between government departments which accelerates the reformation trend towards whole-of-government. A netted environment and a flattened society require greater cooperation between different public departments, leveraging resources without overlapping, and providing seamless rather fragmented public services driven by public demand. Developments in data collection and processing technologies can provide the information basis for lower coordination costs between departments and make further integration of departments possible. For example, in the US, the Department of Transportation and the Department of Public Safety were two separate departments, but after a data analysis revealed a high degree of overlap between the location and timing of public safety affairs and traffic accidents, the two departments began joint policing and achieved good results.

Secondly, there was a further deepening of the governance model in which government, market, and society work together. The concept of governance emerged from a series of problems that arose in practice with centralized and bureaucratic means of management. Governance, compared to management, has a richer connotation, focusing on both the use of governance mechanisms and the role of informal, non-governmental mechanisms when providing public goods. From management to governance, the status of a single government shifted to a new model of multiple public goods supplying. In the era of big data, some enterprises have moved ahead of the government in terms of data storage and mining. Enterprises have a natural sensitivity to the utilization of big data, one example of which is precise marketing based on consumer data and credit card data mining. Google's use of big data analysis to predict flu outbreaks is also a typical example of how companies help in public services with big data. Society is also playing an increasing role in the use of big data in public management. For example, a Washington-based public benefit organization has unified all US federal government expense data on one website, allowing the public to better access and monitor the federal government's expenses and budget. This kind of monitoring from social forces constitutes a powerful regulation of government actions.

3 Openness and Application of Government Data

The earliest application of big data in China was in the field of the Internet, but now it is gradually extending and penetrating into various fields such as finance, telecommunications, industry, and even government, and has generated huge social value and industrial prospects.

The data held by government departments and the public sectors are divided into four types: the first is performance data, which is generally generated and collected in the course of work processes, such as policy documents, specific measures, working meetings, administrative services, etc.; The second is statistical survey data, which is statistics, surveys, and aggregate data conducted by statistical-related industry departments, while the rate of ratification of core government operations, and the degree of digitization rate of data usage should be emphasized; the third is environmental data, which is more often collected through physical facilities, such as cameras installed on the road, coils, water temperature testers, and data collected from other sources, with characteristics as real-time and unstructured like audio and video; the fourth is Internet data, which is the social opinion data that has aroused attentions and discussions of the media and the people about the government's functions or during performing duties, while modern public administration with a basic standard of people's satisfaction, should focus on this type of data. These four types of data constitute big data of government affairs, and their characteristics are very obvious: large volume, wide variety, greater value, but strong monopoly and more difficult to obtain.

There are many applications of big data in government management, for example, public opinion monitoring, performance assessment, industrial planning, integrity regulation, approval services, image management, reputation research, intelligent transportation, disease management, and etc. Some local governments have a lot of interaction with citizens and enterprises when they carry out approval services, especially government call centers. By digging deeper into these service data and analyzing new and valuable things, we can provide better and more refined services to better understand what citizens think and need. Big data governance based on the traditional model, also requires the introduction of new knowledge and technology to complement, integrate, and process each other to form a win–win model.

The application of big data government requires an open sharing mechanism for government data, where precedent is achieved in this regard. Beijing's basic informatization is very well developed, with a wide coverage of network systems (including wired and wireless networks) and an informatization coverage rate of over 95% for core government operations. The construction of four basic databases, namely, the basic information database of legal entities, the basic information database of population, the macroeconomic information database, the basic information database of natural resources and spatial geography, as well as the theme database of each industry have been basically completed, with more than 350 information databases being distributed in the key areas of each industry. In 2011, Beijing completed the emergency IoT project, which is the infrastructure of the application of big data;

while a 6,000 m^2 government cloud room has been planned for the Liuliqiao Bridge area of Beijing.

Previously, when the leaders checked about the number and information about local small and middle enterprises (SMEs) in Beijing, the Commission of Economy and Informatization, the Bureau of Industry and Commerce, and some local taxation, customs, statistics and other departments could only report according to the statistics as "different level of statistics under different statistical standards". The inconsistency in statistical standards and the disparity in results should be a problem for all local governments. What Beijing is doing is to change this approach, from sharing and exchange to integration and sharing. When conducting government information application, enhance the collection data sources diversify which complement each other so that to achieve more authoritative and unified. In 2012, the Beijing Municipal Government Data Resource Portal (data.beijing.gov.cn) was established. This is a government data website led by the Beijing Municipal Commission of Economy and Informatization, with the participation of all government departments in Beijing, which is open to enterprises, individuals and developers in society. The website mainly provides data download services, raw data, geographic information data, online services, etc., with the aim of collaboratively using government data that is scattered across various units.

The application of government big data also has to go through a process of assetization, which has already started in enterprises. There were two big data trading platform set up in Beijing Zhongguancun, but in terms of asset-based management, trading does not represent the whole of asset-based management. If the three issues of personal information protection, government information sharing and opening, and asset-based management are not solved, there will not be a good growth environment for the application of government big data.

At present in China, the national and local governments have successively introduced some relevant policies on data management, which have made some guiding norms on the management and use of government data. For example, data involving people and enterprises cannot be used casually and needs to comply with the relevant laws, regulations, and rules. The use and development of data must be premised on the development of internal protection policies, and only on the basis of protection can it be opened, shared, and provided for use.

In this regard, Guizhou, as a less economically developed province in China, is well aware and has already started action. According to the 2014 GDP data, Guizhou ranks lower than Xinjiang and slightly higher than Qinghai and Tibet, in the bottom six of 31 provinces, cities, and autonomous regions. Due to geographical and transportation factors, Guizhou has been lagging behind in economic competition with other places. However, Guizhou is now betting on big data. In February 2014, the Guizhou provincial government issued and implemented the "Outline of Guizhou Provincial Big Data Industry Development and Application Plan (2014–2020)", marking the beginning of the province's most important strategy for economic development, and becoming the first provincial government to open up government data. In 2014, Guizhou Province launched and completed the construction of the "Guizhou-Cloud Big Data" platform, which is the first cloud service platform in China to achieve data integration, management, and sharing among provincial governments, enterprises, and institutions. In the first Guizhou-Cloud Big Data Business Model Competition, Guizhou opened up the data catalogs of seven cloud systems, including e-government, smart transportation, smart tourism, industry, food safety, environmental protection, and e-commerce, as well as some traffic data of Guiyang City for participants to innovate their business models. From data integration to data application, from livelihood services to business services, the competition received more than 10,000 project candidates from more than 10 countries and regions, including the US, France, and Germany, and more than 20 provinces and cities, including Beijing, Shanghai, and Guangdong, and many universities in China including Tsinghua, covering education, finance, agriculture, medical care, tourism, public services, and other application areas, achieving good results.

In January 2014, Guian New Area was approved as one of the five key new areas set up for the country's western development, and in February 2015 it was approved as the country's first demonstration area for the development of big data industry, which has already attracted the three major operators Mobile, Unicom, and Telecom to set up. These three major operators planned to invest over RMB 15 billion in the construction of national data centers; in addition, Foxconn, JD, and Alibaba also have set up related projects in Guizhou.

It has taken 18 years for government data to be made public in the US. It is believed that China should not have to wait too long, as this is the prevailing trend.

4 Breaking Information Silos and Upgrading Livelihood Services

In terms of big data, the more relevant it is, the more valuable it is; the more open it is, the more valuable it is. The government should take the lead in opening up public data such as administrative records and public resources in the public domain generated in the course of government governance, encourage various institutions as well as other non-government and non-profit organizations to provide business data generated in the course of their public services, and strive to promote for-profit organizations such as enterprises to open up the relevant data formed in the course of their production and operation, network transactions, and etc. The various types of data resources needed in the process of economic and social development require greater openness and sharing of data, and establishing organizations for data convention in the public service sector, based on the continuous strengthening of government information disclosure. In recent years, we are pleased to see that some local governments, ministries and agencies, enterprises, and institutions have started to make a breakthrough in data opening and sharing. We believe that in the near future, as data management becomes increasingly standardized, data privacy protection measures become better and better, and technical capabilities become bigger and bigger, more value will be revealed from data sharing and opening.

Starting with the breaking down of information silos, deeply carry out the reform of integrated, grid-based, and information-based service management, sorting out social problems highlighted in the urbanization process, such as the sense of social security, the satisfaction of the public, the happiness of the public, the health index of the residents, the management of the migrant population, and the hidden risk of social security, into a comprehensive data, and then making decisions to form a public service pattern of big data integration, big data service, and big data management.

There are many successful examples of the application of big data to government public services. For example, Beijing "12,345" handy call center serving as a comprehensive reception and dispatching platform for non-emergency assistance services in Beijing, can integrate the city's handy call services, support all kinds of requests, demands, criticisms, and suggestions from the public, and provide convenient and fast public information services. Another example is the "one center and four platforms" (the city big data center and four comprehensive information service platforms for e-government, city management, economic operation, and livelihood services) officially launched in Wuxi, Jiangsu Province in 2015, which have brought together 5,799 data items and more than 140 million pieces of data from 34 commissions and departments in Wuxi. This has basically built a basic information database for population and legal entities, realizing real-time presentation and data mining and analysis of the operation of the four major areas of e-government, city management, economic operation, and livelihood services, and initially forming a big data ecosystem of "unified deployment of the city's governmental data, a unified gathering of foundation data, in-depth integration of business data, in-depth mining of application data, sharing of subject data across regions, sectors, and levels, and comprehensive improvement of the directory and exchange system". In particular, the "Citizens Web Page" built based on these platforms, creates a personalized portal for each Wuxi citizen through aggregating information and services that are closely related to citizens and enterprises, providing functions such as government information subscription, information and service customization, and government-citizen interaction. At present, the Citizens Web Page can provide access to 958 public services at the municipal level, and has achieved online access and processing of information on 24 public services, including human resources and social security and provident funds.[2]

There are many other cases like Beijing and Wuxi where the government has used big data to improve public services. So, as the migrant population continues to change and increase, can these services provided by the government be made available to everyone, including the migrants? A lot of interesting things happen when a database of information on population, legal entities (institutions), and housing is created. For example, a comprehensive analysis of home addresses and work addresses will show you that there is a community like Beijing's Tiantongyuan, where a particularly large number of people live, but is particularly far from the city center. Many of these people have to go to work in places like China World. This is when the government has to think about whether it should add a few more bus lines between these two locations.

---

[2] LIU Chun. Wuxi City Big Data Center Launch [N]. Wuxi Daily, 2015-12-26.

How the lines should be designed and how often should they depart, this can all be worked out from the data.

In many cities, the elders can now take public transport for free, which is seen as a socially progressive move, but you still have to present your "senior card" when you get on the bus, and only local citizens with household registration can apply for a senior card. At the Shenzhen Intelligent Transport Command Center, we were told that elderly people over the age of 65, whether they are migrant residents or tourists, can all take public transport for free by swiping their identity cards, which is seen as equalization of public service. Another example is that Shenzhen has set up a large number of community health centers in each neighborhood, which are required to provide public health services with "the same services, the same standards, the same protection, and the same assessment" to both the household and migrant populations. In the integrated system of information on population, legal entities, and housing, you can see the distribution and structure of the city's population, as well as the distribution of the city's community health centers, where medical resources are too concentrated and where they are relatively scarce. However, the so-called "equalization" is not "average", and it is not just random like sprinkling a handful of salt, but also a reasonable and effective allocation of health care resources. For example, the staffing and equipment allocation of community rehabilitation centers near industrial parks should focus on family planning and pedestrian volume; while in communities with a high concentration of elderly people, there is a great demand for the treatment of chronic diseases.[3]

The above case shows us that if we simplify public services to a formula, then its denominator is population and the numerator is public services. The first step to equalization is to find out how many people there are and how many public services the government can provide in the present and future. It is not enough to know only the number of people, but also the structure of the population, how many elderly people and school-age children there are, how many registered and temporary residents there are, as well as how many people came for business and tourism. All this can be told by big data.

5   Big Data Supports the Equalization of Public Services

The goal of achieving overall prosperity and equalization of basic public services by 2020 is the main measurement marker. The State Council's 13th Five-Year Plan for Promoting the Equalization of Basic Public Services pointed out that "China's basic public services still have shortcomings such as insufficient scale, poor quality, and unbalanced development, highlighted by: unbalanced resource allocation between urban and rural areas, uncoordinated hardware and software, and large differences

---

[3] SUN Xiaoli. Optimization of Public Services in the Context of the Era of Big Data [C] // Proceedings of the Cross-Straits Seminar on "Administrative Reform and Modernization of Public Governance". 2015.

in service levels; Inadequate and underutilized fundamental facilities coexist, and a serious shortage of talents; blind areas of coverage existed in some service programs and without effectively benefiting the entire mobile population and groups in need; lagging innovation in institutional mechanisms and insufficient social forces participation." Data from the National Bureau of Statistics also shows that although the construction of equalization of basic public services in China has achieved positive results and the situation of equalization has been improved, the level of balanced development still requires to be further enhanced.

Compulsory education, public health, basic medical care, basic social security, and public employment services are the most concerned and urgent public services for urban and rural residents. Big data technology can not only help the government's public service decision-making and supervision but also provide personalized and precise services to the public, as well as support the government in providing "equalized public services". The US is a world leader in the application of big data to the management of public health. In the US, data on the quality of healthcare and performance of different hospitals is made public by the government. This not only helps hospitals to improve the quality of healthcare service but also saves on healthcare costs. According to resources, the Clinical Decision Support System alone can reduce US healthcare expenses by \$165 billion a year.

On 31 August 2015, the State Council released the Action Outline for Promoting the Development of Big Data, of which the main related to medical big data is the open sharing of government data. The main tasks include: building a database of electronic health records and electronic medical records to construct a big data application system for medical and health management and services covering public health, medical services, medical security, medicine supply, birth control, and comprehensive management services; exploring services such as booking, graded diagnosis and treatment, telemedicine, sharing of examination and test results, combining prevention and treatment, integrated eldercare with medical care, and health consultation to optimize the formation of standardized, shared, and mutually trusting treatment processes; encouraging relevant enterprises and institutions to carry out research on the innovative application of medical and health care big data to build integrated health service applications, and etc.[4] Zhengzhou Municipal Bureau of Health is making every effort to build a "medical and health informatization city", establishing a regional integrated health information service platform with "one exchange platform, one data center, and nine application systems" covering both urban and rural areas to provide the public with open, transparent, efficient, convenient, high-quality, and equalized public health and basic medical services. The platform makes full use of big data thinking to collect information on 7,809 personnel and objects in 12 categories who combating 198 responsible areas, public areas, and medical institutions for the practice of medicine without a license, realizing information interchange and resource sharing among health administration departments, medical and health institutions, and grassroots health service units, mainly addressing real-life problems

---

[4] Excerpts from the Notice of the State Council on the Issuance of Action Outline for Promoting the Development of Big Data.

including the public's needs for information on medical and health services, communication between doctor and patient, prevention and control for infectious diseases, emergencies, and basic medical services, and etc.[5]

Big data is equally important for education, as it can contribute positively to the achievement of educational equalization. At the 2016 Education Big Data Expert Forum, YAN Jianhui, Director of China Statistical Information Service Center, pointed out that big data has unlimited potential to reshape education. China needs a large number of educational big data researchers and practitioners to give full play to their creativity and combine advanced technologies such as data mining, learning analysis, artificial intelligence, and visualization with modern educational reality issues. The ultimate value of education big data should be reflected in the deep integration of mainstream education business as well as the transformation of think tank education. Achieve education management scientification, education model reform, and personalized learning driving, to truly drive education evaluation system restructuring, drive scientific research paradigm transformation, and drive education service humanization. In the current situation, education big data should be strategically defined as an innovative strategic asset to drive educational change, a scientific concept to promote comprehensive reform in the field of education, and a cornerstone for the development of think tank education.

In this seminar, LIU Haimin, former Vice President of Northeast Normal University, summarized the value of big data from four aspects: firstly, it is used to illustrate reality; secondly, it is used to show social concerns and value trends; thirdly, it is used to help the subjects to understand themselves; and fourthly, it is used to reflect the development of education. He believes that from a future perspective, leaving this data asset behind is of great importance in understanding historical research, the historical lineage, historical evolution, and the state of development at that time of China's educational development.

Big data can also be very useful for the development of the cultural industry. Through big data technology, data analysis of most people's preferences can clarify the tastes and needs of the target audience and create marketable cultural products, thus reaching the widest possible consumer base with reduced risk. The use of big data can effectively extend the industrial chain of cultural products and increase the profitability of cultural enterprises. For example, the film and television industry can discover the most popular film and television derivatives through deeper mining of audience comments, carry out precise secondary product development, and even potentially develop new business models. The results of big data analysis techniques in historical and cultural analysis research are important for us to further deepen our understanding of the history of the development of Chinese civilization, to recognize and identify the "genes" of Chinese culture, to continue the cultural lineage, to clarify the key elements of our cultural construction that should be strengthened for protection, inheritance and external communication, and to formulate national cultural development strategies. The third census of cultural relics shows that there

---

[5] WEN Youcheng. Zhengzhou Medical and Health Services Enter the Era of Big Data [N]. Economic Information, 2014-09-26.

are 766,722 pieces of immovable cultural relics in total, and is estimated to be in the range of 300 million pieces of movable cultural relics. Many museums, libraries, organizations, and departments for the preservation of intangible cultural heritage, are currently working on the digitization of various historical and cultural resources in different ways and for different application purposes, objectively creating an unprecedented and rare collection of Chinese cultural resources.[6]

In the field of public cultural services, where demand is the top concern, big data can directly serve the development and promotion of public cultural products and provide good support and assistance for the construction of digital cultural centers. Currently, the more realistic and operational is the construction of a digital experience hall. It requires certain technical support: firstly, it relies on the backstage of the cultural museum's website as a resource library to transmit the latest service contents; secondly, it relies on a variety of digital application modules to experience various art products and services of the cultural museum; thirdly, it relies on digital mean to present cultural scenes with local characteristics in a panoramic manner in the experience hall and create an immersive audio-visual experience for users; fourthly, it relies on connecting the television screen, computer screen, tablet screen, and mobile phone screen, so that the contents are interchangeable and interactive, and the display contents can be zoomed, played on demand, touched, moved, and downloaded to create a new cultural experience.[7]

6 Big Data Enhances the Effectiveness of the Construction of Digital China

The development and application of big data must be a "managerial level project" because it requires the mobilization of resources from all sides. Big data technology is not the most difficult as analysts and researchers are both available, the key is whether you utilize these people and how to utilize them, whether to have the thinking doing this thing. This requires the overall leadership that with only one responsible person from a specific department it will be very difficult to coordinate and promote.

The real value of data is its ability to scan the current situation, assess its effectiveness, and provide insights into the inner pattern of government or business management. Big data can be mined and analyzed to help leaders make scientific decisions and to explore possible solutions for government public management. Big data can also be a "booster" for organizational effectiveness.

Big data enables government departments to better understand the thoughts and needs of the public and to visually identify and address the most significant issues that are most strongly reflected in public opinion; it also allows the public to clearly see

---

[6] JIANG Nianyun, TENG Jipu. Using Big Data Technology to Promote Cultural Resource Management [N]. Science and Technology Daily, 2013-10-11.

[7] ZHENG Zhiyue. The Construction of Digital Cultural Museums in the Perspective of Big Data [N]. China Culture Daily, 2015-05-15.

the whole process of government operations, which is conducive to better monitoring of government administration and building a transparent government. Therefore, in the era of big data, leaders must advance with the times, actively learn and make good use of various modern information processing technologies rather than rejecting the challenge of new things to traditional experience and applying them skillfully in their work to better serve society.

Big data mining and analyzing can provide effective help to organizations in scientific decision-making, thus enhancing the ability to predict economic and social development and exploring feasible solutions for government public management as well as the sustainable development of modern enterprises. Sharing and integration is the only way to generate value, and the promotion of big data makes data sharing possible. China already has a good foundation for informatization, and the existing databases retained by the government in the process of social governance can be efficiently interconnected, breaking through the bottleneck of information silos and greatly improving the ability to collaborate between government departments. Big data applications can not only improve the efficiency of serving the people, as "if data does more, then the people can do less", but also reduce government management costs. In addition, most important, big data applications can also provide strong support for government decision-making, and the wisdom from the data will effectively promote the overall level of government management.

On the issue of supporting economic and social development, enterprises use big data to maximize profits, government departments use big data to formulate scientific policies, and academics use big data to find scientific laws. Big data is not only a vehicle for creating value, it affects various aspects such as city management, e-government, public opinion monitoring, enterprise management, etc. Once one has mastered the application of big data, the development dilemma of the smart city may be easily solved.

Since 2013, the National Bureau of Statistics (NBS) has been promoting the application of big data in government statistics, and MA Jiantang, Ph.D. in economics and director of the NBS at that time, said "whoever owns big data will take the strategic high point. Big data will definitely become the information basis for macro-control, national governance, and social management for the government". The development of information technology has facilitated the collection, analysis, and dissemination of data by government statistical departments. Big data has deeply involved in all aspects of society.

The following two case studies are presented to deepen the understanding of the application of big data in the construction of Digital China.

Case One

On 6 August 2015, the Foshan Daily published an article "Making the Best Use of Big Data, Improving Efficiency and Facilitating the Development", stating

Big Data Promotes the Construction of Digital China

that Gaoming district, through the "One Entrance" comprehensive law enforcement platform, has incorporated 31 functional departments in the district into the platform for unified management, which has comprehensively standardized and improved the level of administrative enforcement. The model has been affirmed by a lot of experts and scholars from the National Academy of Governance, Guangzhou Academy of Social Sciences, and Sun Yat-sen University, saying that it has great demonstration and promotion value.

Reporting with the "One Entrance" platform is more convenient and effective. The public can simply call the 12,319 Complaint Hotline or log on to the Gaoming District "One Entrance" comprehensive law enforcement website if they have something to report. Once the report enters the platform, it will be automatically assigned according to the legal authority of each department to achieve unified management, unified transfer, unified supervision, etc. The convenient model for reporting from the public has gained the attention of experts and scholars. PENG Peng, a senior researcher at the Guangzhou Academy of Social Sciences, appreciated and said that the purpose of the government hotline was to facilitate the public to call, but in the past, each department has its own hotline being confused for the public to report. The unified 12,319 Complaints Hotline in Gaoming district makes it convenient for the public. Professor YANG Xiaojun, deputy director of the Department of Law of the National Academy of Governance, believes that this model is actually an amplified application of the "110" model, and it is a very good attempt for Gaoming District to amplify the model to the entire government law enforcement level. Closed management on reports from the public, on one side, making it convenient for the public, reflecting the function of service-oriented government, and on the other side eliminating the current administrative law enforcement inaction, slow action, and selective action. The unified platform supervision is conducive to standardized administration and improves the ability of grassroots self-monitoring.

Once checks and balances can be monitored online, justice can be shown to the public. Gaoming District's comprehensive law enforcement reform does not change the framework of the original administrative unit but rather uses the "Internet+" model to rebuild the workflow, ensure online supervision of authority, and break the barriers in the interface between administrative law enforcement and criminal justice. ZHANG Xiaoyu, associate professor of the Law Department of the National Academy of Governance, commented that the comprehensive administrative law enforcement reform he knew before was mainly about institutional integration, while the "One Entrance" comprehensive law enforcement reform in Gaoming District was about integrating information. It is not easy to bring all 31 administrative and law enforcement units into one platform, especially when public security is also brought in with an early intervention mechanism for public security implanted. DING Li, deputy dean of the law school at Sun Yat-sen University, said: "The reform in Gaoming District did not change the original framework of administrative units, instead changing the structure of the relationship between government departments and the rules of the game through a seemingly simple process re-engineering, and the corresponding balance of behavior has changed as well. This reform has, on the one hand, made information no longer exclusive to one department, and on the other hand, the actions of law enforcement agencies are known to the supervisory organizations, which has led to more restrictions on the opportunistic behavior of many departments who use hidden information and actions for small-scale defense or even personal interests."

Reform with the utilization of big data is more effective. For the follow-up development of Gaoming District's "One Entrance" comprehensive law enforcement platform, PENG Peng suggested that in the era of big data, the advantage of "One Entrance" for report should be used for statistics to understand which departments in the region have a high concentration of complaints, the frequent complaints topics, and regional distribution density. In terms of the analysis of big data, ZHANG Xiaoyu also put forward the same advice that "this information should be used to strengthen the analysis and research, summarize the key characteristics of the public report, and make it the information guidelines for government to improve their work." He also pointed out that the principle of this reform is not complicated, and the promotion of it depends mainly on the coordination ability of the local party and government. After all, many departments of our local government are deeply influenced by rules and regulations, and many of them belong to vertical management departments, which are not easy to integrate. Although it might be difficult, the reform in Gaoming District is exemplary and worth promoting.

Case Two

On 1 October 2014, the Qingdao Urban Intelligent Traffic Management Service System was completed and launched. This service system contains one data center, three major platforms for information management, comprehensive application, and intelligent management, as well as eight sub-systems for signal control, traffic flow collection, traffic guidance, information dissemination, comprehensive monitoring, traffic enforcement, command and dispatch, and security management.

According to the Qingdao Morning News, the city's intelligent traffic service system used big data technology to conduct in-depth mining and analysis of more than 18,000,000 pieces of data collected daily, accurately identifying traffic congestion hotspots and areas, providing scientific and accurate data basis for traffic organization optimization, signal control optimization, suspect vehicle tracking, and other works. At present, the system has reached 6.2 billion records of passing vehicles, and finds any vehicle license plate within 5 s. Tested on tens of billions of data, the search time can be controlled within 10 s, reaching an international leading level. At the same time, with the use of intelligent analysis, it can achieve automatically alerting within 5 min, reducing the detection time by 75%. In accordance with the principle of "police force following the police information, flat command, and grid management", the urban area was divided into 55 police districts, and a model of "point and patrol combination with patrol-oriented" was built at the same time to realize quick detection, quick response, and quick disposal of police situations. This enables quick detection, quick response, and quick disposal of police information. Statistics show that in the first year, Qingdao's intelligent traffic system has reduced the number of congestion points in the city by nearly 80%, the congestion level in the morning and evening peaks in the whole city has been significantly reduced, and the number of stops for vehicles has been reduced by 45%, and the average travel time of main roads has been shortened by 20%. Among them, peak hour congestion mileage of the key road Shandong Road has been reduced by about 30%, while the passage time has been reduced by 25%. Since the system has been in operation, the overall average speed of the urban road network has been increased by approximately 9.71% and the travel time has been reduced by approximately 25%.

Governance Modernization and Governance Digitization

Qing Jiang

1 Big Data Offers New Ideas for Social Governance

Saying goodbye to the "captain's call" and using facts to make decisions is not only an appeal by experts but also one of the "guidelines of governing the country" that has been repeatedly stressed by the top leadership of the State Council in recent years. On 5 March 2015, Premier LI Keqiang said, "Great truths are always simple, and power cannot be used as one pleases" in his Government Work Report at the third session of the 12th National People's Congress. On 17 June, Premier Li Keqiang stressed at the State Council's executive meeting that "the utilization of big data and other modern information technology is an effective means to promote the transformation of government functions, streamline administration, delegate power, improve regulation, and upgrade services." At the same time, it was clearly pointed out that the datafication of decision-making based on the application of big data is an important element of government reform and innovation.

Professor LI Liangrong, Director of the Center for Communication and State Governance Research at the Fudan Development Institute, has also pointed out that: "The unified will of the government and the plurality of demands between different sectors of society are the focus of national governance." It is necessary to effectively and comprehensively understand the dynamics of public opinion and combine it with the will of the government in order to form the "greatest common divisor" in society and form the basis for policy formulation.

In the past, methods of obtaining public opinion such as holding seminars and conducting public opinion surveys had the problems of small coverage, poor timeliness, and poor feedback channels, creating "intermediate obstructions" and high costs. In the era of the Internet, the use of modern information technology such as big data has opened up new ideas for national governance. LI Liangrong said, from

Q. Jiang (✉)
Zhongnan University of Economics and Law, Wuhan, Hubei, China
e-mail: 18712919290@163.com

© China Renmin University Press 2023
Q. Jiang (ed.), *Digital China: Big Data and Government Managerial Decision*,
https://doi.org/10.1007/978-981-19-9715-0_9

a global perspective, that though e-commerce, Internet finance, and technological innovation are developing rapidly, it is just the beginning for government to utilize big data.

"We hope that through the deep mining of data, we can understand the real situation of public sentiment at a lower cost so that the multiple demands of society can be integrated into the government's governance framework and form a more effective governance framework." LI Liangrong said that by using the Internet as a new tool to measure, collect, and convert data, with further improving methods and algorithms for deeper mining, a structural and logical understanding of the issues can be realized, based on which the multiple frameworks and multiple issues can be effectively combined into an organism with complete superstructure and active sub-foundation.

Professor WANG Jun, President of the Guangdong Academy of Social Sciences, pointed out that three features of social governance have emerged: first is a plurality of interests, as traditional social management is a dual structure of the supervisor and the supervisee, while modern social governance is based on a variety of interests, such as government, enterprises and institutions, social organizations, trade and community organizations, and etc.; second is the widespread use of big data technology; leading to the final third characteristic that social governance has become a government-led cooperative co-management. WANG Jun said that traditional social management has not adapted to the needs of development in the new era, but the new management model has not yet been established. In order to promote smoothly in an organized and orderly manner, it is necessary to explore a new model of cooperative and shared governance under the leadership of the government, with the participation of multiple parties that put the people first.

Professor SHEN Guchao, Deputy Director of the Collaborative Innovation Center of South China Sea Studies at Nanjing University, gave a presentation on "How to Improve the International Discourse of China's Public Opinion in the Era of Big Data", taking the example of the maritime territory rights protection. He suggested that facing massive amounts of data, we must learn to use technology to transform data into medium and small data and apply it to the practice of improving international discourse. SHEN Guchao believes that constructing the evidence chain in the maritime territory rights protection, of which the concept of all-source intelligence has emerged abroad, is actually equivalent to the "big data", linking five databases: image database, document database, legal database, map database, and dynamic database. Therefore, he suggests that building international discourse should start from three aspects: public opinion, big data, and evidence chain turning big data into small data that users can process and extracting intelligence and knowledge from small data that serves decision-making.

Using the data aggregation and data analysis capabilities of big data frameworks to achieve innovation in social service management work can significantly enhance the comprehensive capabilities of grid-based management and social services. Gather the foundation data of government departments, expanding social data through a dynamic public opinion collection system, resident's feedback log, and event record data resources in streets and communities, to effectively cover streets, communities, and grids; in the data aggregation level, a multi-dimensional database of people, places, events, things, situations, organizations, and houses is assembled to enable the interconnection of indicator data; in the data analysis level, the three main lines of grid-based social management, social services, and social participation are analyzed using statistical analysis and mining analysis techniques, focusing on the patterns and characteristics behind the data on human management and services; in the capacity enhancement layer, government departments realize linkage and sharing based on data analysis results, and the government and residents realize interaction and communication based on the data sharing service platform.[1]

With the associated technologies of big data framework, it is possible to identify the public service needs of different groups of people during the economic and social transition period, optimize the allocation of workforces, improve the efficiency of departments, modify primary-level government management, and increase public satisfaction.

Using the data analysis technology of big data framework, the Beijing Dongcheng District Integrated Service Center for Social Affairs Management regularly compiles statistics on the requests of key service groups on the Social Affairs Management Information Platform, tracking and analyzing the handling of these requests to identify recent affairs with a high concentration of public requests and focus on manpower to resolve them. For example, at the end of 2012, when the weather became colder, the statistical analysis results of the Social Affairs Management Information Platform showed that the public sentiment and feedback related to heating has increased significantly. Based on this situation, the relevant departments in Dongcheng District organized special activities in a timely manner to investigate heating problems in key areas, effectively preventing and solving all types of potential fire hazards due to the heating supply.[2]

The above cases fully demonstrate that the era of big data has placed new demands on government governance. The government needs to have comprehensive, fast, and timely information, filter invalid, and misleading information, use logical and data analysis methods to disclose and share information in a timely manner, and conduct an in-depth analysis of typical cases in its areas of concern to form a case base for guiding its future work.

With the advent of the era of big data, whether for government or enterprise, the application of big data management decisions is the trend, leading to the final direction. We must understand that in the era of big data, the leadership of the "captain's call" can be ended!

---

[1] HUANG Shaohong, XIE Songxin. The Government Uses Big Data to Say Goodbye to "Captain's Call" in decision-making [N]. South Daily, 2015-07-27.

[2] CHEN Zhichang. Research on Using Big Data to Promote Governance Capacity Modernization— Illustrated by the Case of Dongcheng [J]. Chinese Public Administration, 2015(2): 38–42.

2 Digital Regulation to Correct Market Distortions

First of all, it is important to know what exactly the regulation is. Breaking the regulation down, it includes supervision and management. Government regulation is primarily embodied in the marketplace. The essence of market regulation is to correct market distortions, to correct the problems that arise in the market so that it can continue to move forward according to the conventional way of operation. The traditional way of regulation normally is to issue various kinds of licenses such as industry and commerce, quality inspection, food and drug, taxation, environmental protection, health, safety, and so on, as well as impose fines. Licensing is the first step, followed by various inspections, if during which problems are found, penalties are imposed, in the form of fines, closures, rectification, etc.

This traditional way of regulation is now challenged by the fact that it is impossible to find anyone to issue a citation. Why? Because the Internet has developed and e-commerce transactions have emerged that you cannot find an object or proof in the traditional shops for imposing fines, as the transaction is no longer generated in a physical storefront. Therefore, many departments are now starting to consider that the tax authorities should care about how many transactions are actually happening online, how large the transactions are, and what the structure of the transactions is. This is a challenge that comes from information technology and every regulatory authority is currently experiencing the same problem of having to face a change in the way they regulate due to the prevalence of e-commerce. Although e-commerce is not the whole of society, it is a fact that it has become an important way of human life and will continue to develop.

Innovation is the generation of new solutions to problems within existing ways of thinking. The work of an administrative unit keeps the same almost every year and its functions are relatively stable, so this "stability" requires innovation in thinking and some changes in problem-solving.

How can big data be applied to achieve regulatory innovation? The National Medical Products Administration which has utilized big data as tools for food safety regulation and achieved remarkable results is a good example.

It may be noticed that the image of food safety in China has gradually improved in recent years. Why? A study conducted by the China Statistical Information Service Center during the third session of the 12th National People's Congress and the third session of the 12th National Committee of the Chinese People's Political Consultative Conference (CPPCC) in 2015, found that this problem was at the top of public opinion concerns for two consecutive years in 2012 and 2013, receding to sixth place in 2014, while in 2015 it was no longer in the "top ten" of public opinion concerns. By analyzing the data of media, experts, and the public on various dimensions of causes, we found that government regulators had done a lot of work around food safety, which resulted in the "China Food Safety Internet Image Study Report". As

Governance Modernization and Governance Digitization

we all know, people are self-protective and will not eat something they know is harmful. All the unsafe incidents that happened in the past were consumed because people were unaware of these hidden dangers. And in the past, food safety supervision was more of an afterthought, as was the case with commodities and consumer goods. The work done by the departments of industry and commerce, quality inspection, and food and drug were basically afterthought supervision. This is the traditional way of regulation, which is also a common problem caused by "information asymmetry". Now as it is an information society where technological means are more abundant, shouldn't regulation move forward? If you have enough information, it is possible to move the regulation forward. One of the important things that the National Medical Products Administration is doing now is to make the data on product sampling public and to constantly publish the results of product sampling on the official website. As long as people have the awareness of self-protection, they will make reasonable and safe consumption based on this information. This is the positive effect of information symmetry.

Upholding information symmetry is a fundamental focus of regulatory innovation, and many problems will be solved when information is symmetrical and communication is smooth. It is believed that in the future, the national food and drug department, industry and commerce department, quality inspection department, and other departments related to people's livelihood will develop a lot of App, mini-programs, H5, and other mobile convenient forms, in order to allow people to find out the information they need anytime and anywhere on their mobile phones, making life more convenient and better.

3 Big Data Enhances the Delicate Level of Government Governance

Big data brings the possibility of delicacy management and new opportunities for service innovation for governments and enterprises while flattening has become a characteristic of management in the era of big data.

There are two forms of organizational structure, namely flat structure, and hierarchical structure, according to the relationship between management layers and management span. A flat structure is a structure with a few layers of management and a wide management span, while a hierarchical structure is a structure with multiple layers of management and a narrow management span. With a narrow span, the number of management layers and managers has to increase, as well as the amount of effort, time, and costs; whereas with a wider management range, the number of management layers can be reduced, as well as the number of managers, time, and costs required, and the work efficiency can be improved with less transferring channels of information between the superior and lower levels.[3]

---

[3] ZHANG Xiaoxia. The Application of Flat Management in Enterprise Management [J]. China Logistics and Purchasing, 2005(8).

Flat management is a management model as opposed to a hierarchical management structure, which better solves the disadvantages of hierarchical management such as overlapping layers, redundant staff, and inefficient operation of the organization, and speeds up the rate of information flow.

According to the South Daily on 27 July 2015, the first "Guangzhou New Observation" roundtable meeting was held at the Administration Building of Jinan University, where experts and scholars from the fields of politics, academia, research, and enterprises, as well as media reporters, discussed the innovation of government governance in the era of big data. The experts pointed out that big data will allow the government to say goodbye to "captain's call" decisions and make government decisions more scientific and delicate.

JIANG Shuzhuo, Secretary of the Party Committee of Jinan University, said that big data provides a solid foundation for social governance and government governance, providing not only the concept of governance but also a macro belief making our decisions more solid and evidence-based, which is no longer making just captain's call decisions. We are now making decisions based on data and facts in order to get the governance model right.

The roundtable meeting discussed how big data can be used to solve the problems faced by the Guangzhou government and help improve and innovate government services in the city. In his speech, ZENG Weiyu, President of the Guangzhou Federation of Social Sciences, pointed out that the purpose of the "Guangzhou New Observation" series of seminars is to serve the scientific decision-making of the Party Committee and the government in the light of Guangzhou's development practice. At present, Guangzhou is actively implementing the national "Belt and Road" initiative, the construction of the Pilot Free Trade Zone, and the innovation-driven strategy, while accelerating the development of an international shipping center, a logistics center, a trade center, and a modern financial services system. There are a large number of decisions that need to be consulted in this process, a large number of questions that need to be answered by the academic community, and a large number of policies that need to be innovated. All these cannot be done without the collection and analysis of big data.

YAO Yanyong, a member of the editorial board of the South Daily, pointed out that in the era when everyone has a microphone to shout out, society is in greater need of rational and constructive voices. At present, China's reform is entering a critical phase fraught with tough challenges, and Guangzhou, as a national central city, is also ushering in a new round of opportunities and challenges for reform and development, which urgently requires the participation of various academic circles. He said that "We expect these series of seminars will attract more experts and scholars to participate, give full play to the role of the think tank, making appraisal or critical advice and suggestions based on theory or practice around the reform and development, difficult, common, and focused problems in grassroots governance, to further develop reform and development consensus, thus promote the democratization and scientification of government decision-making."[4]

---

[4] HHUANG Shaohong, XIE Songxin. The Government Uses Big Data to Say Goodbye to "Captain's Call" in decision-making [N]. South Daily, 2015-07-27.

Data Security and Risk Management

Qing Jiang

1 The Challenges for Government Applying Big Data

The government also faces the following particular challenges when applying big data.

1. **Data Collection**. As the government collects data from not only different offline departmental channels such as the nation, organizations, and departments, but also from a variety of online channels such as social networks and the Internet, making collection difficult as you can imagine.

2. **Data Sharing**. The difference between commercial data and government data is the range and scope, which have been growing steadily in recent years. The government accumulates a large amount of data in the course of providing public services and enforcing laws and regulations. The characteristics, attributes, values, and changes coming from the storage and analysis of this data are all different from the data generated in the operations of commercial companies. The characteristic attributes of big data in government can be expressed as security, storage, and diversity. Typically, each government agency or department has its own repository for public or confidential information, but there are no uniform standards for this sectoral information, and for various factorial reasons, departments are reluctant to share their proprietary information.[1] This is one of the problems with data openness today: a lack of willingness. In addition, compartmentalization, as well as overcautious and indecision, are also a problem in opening up data in government departments.

---

[1] GAO Changshui, JIANG Daohui. The Advanced Applications of Big Data in National Government are still in the Early Stage [N]. China Electronic News, 2014-05-30.

Q. Jiang (✉)
Zhongnan University of Economics and Law, Wuhan, Hubei, China
e-mail: 18712919290@163.com

© China Renmin University Press 2023
Q. Jiang (ed.), *Digital China: Big Data and Government Managerial Decision*,
https://doi.org/10.1007/978-981-19-9715-0_10

The fact that each government departmental system keeps information that is completely isolated from other departmental systems makes data integration between government organizations and departments even more complex, resulting in the extremely high cost and frequent failure of departments to communicate with each other, which is a major factor affecting data integration. For example, if the police and the hospital could share information on violent crime, crime will be better characterized; however, the key to the success of this case is the ability to communicate smoothly between the two organizations. In addition to the communication factor, we also found that although most government data is structured, collecting data from multiple sources and channels remains a significant challenge. The lack of standardized data formats and software, as well as cross-agency solutions for extracting useful information from the scattered databases of multiple government agencies, are also major challenges for the government in promoting the application of big data.

3. **Legal and Regulatory Safeguards**. Necessary laws and regulations should be defined between the collection and use of big data for predictive analysis and safeguarding citizens' privacy. The USA PATRIOT Act does not only allow legal surveillance but even surveillance of citizens. In contrast, relevant laws and regulations have not yet been formed in China, and there is no clear boundary between citizens' privacy and normal applications of big data. In this case, it cannot be monitored when some enterprises using citizens' privacy to engage in commercial activities, and even government departments have not properly enforced their due responsibilities, while some data that could have been opened up cannot be shared because there is no corresponding boundary. Therefore, it is now imperative to introduce a corresponding act as soon as possible to regulate and facilitate data sharing.

4. **Data Security Protection**. This is probably the most fundamental attribute of all government big data. Although now the country is trying to promote government data openness, however, most big data technologies, including Casandra databases and Hadoop distributed technologies, currently lack adequate security tools. Technical security concerns are also a barrier to data sharing and openness. At the same time, the biggest or most likely trigger of leadership responsibility for government departments in terms of collection, storage, and utilization of data is the security factor. Until there is a clear attitude from the higher authorities, I am afraid that the first consideration of the leaders of the grassroots departments is the issue of their own responsibility based on security, rather than the development possibilities brought by the open sharing of data. With the advent of the era of big data, leaders will also face new management risks and challenges in managing their schedules. The growing economic and social development has led to an increasing amount of structured and unstructured data, which, together with the increasing diversity of analytical tools, has enabled a continuous shift in organizational risk management towards a data-driven approach.

2 Data Security Should Not Be Overlooked

It is risky to open up government data. However, in order for big data to help scientific decision-making and improve public services, it is necessary to promote the interconnection sharing and opening of the data. What we need to do now is to get a clear picture of our database. In Article 20 of the Annex to the Several Decisions on the Utilization of Big Data to Enhance Services and Supervision of Market Subjects, relevant departments are required to explore the establishment of a catalog of government information resources, and a guide to cataloging development is required by the end of December 2016. Once the data is unlocked, the data from all parties will present many new elements like a kaleidoscope.

It is true that it is more than generally difficult to connect government data, and the first is to solve the problem of compartmentalization. Some people are always worried about the "trouble" that might occur after the unlocking, taking the privilege of the so-called voice by withholding the power and waiting for others to come for inquiry. The second is over-cautious and indecisive, like the problem when they always wait for others to start. Why is big data called the "project of the top"? The only way requires the top, the manager to unify their thinking. Now big data is as important as a national strategy, the top is striving at the political level and the foundation is promoting at the technical level, therefore, it is impossible for the middle to be stuck at the application level.

It should be seen that the importance of data security has long been recognized by our government, which has also strengthened associated work in this regard. The sluggishness and blindly following of some officials who care about only superficial performance, like the behavior of some corporate data thieves, has constrained the development of big data in China, which has made these big data practitioners feel very uneasy.

As an increasing number of data is produced online through online interactions and transactions, the motivation for crimes based on data theft is stronger than ever. Today, hackers are more organized, more professional, more powerful in their tools, and more sophisticated in their methods of operation. Big data in the cloud is a very attractive target for hackers to obtain information. Compared to the past when there were just one-off data leaks or minor hacking incidents, now a data leak can result in a serious loss for the whole enterprise, which includes not only damage to reputation, huge economic losses, and, in serious cases, legal liability. Therefore, in the era of big data, network resilience and prevention strategies can be crucial.

But there is still considerable pressure on governments and enterprises to rapidly adopt and implement new technologies such as cloud services, as they can pose unforeseen risks or have unintended consequences, placing greater demands on enterprises and governments to develop secure and correct cloud procurement strategies.

As we all know, the collection, storage, access, and transmission of data must have the help of mobile devices, so the advent of the era of big data has also led to a surge in mobile devices. With this comes the rise of BYOD (Bring Your Own Device), and an increasing number of employees are bringing their own mobile devices to the office. While BYOD has undeniably brought convenience and significant savings to businesses, it has also brought greater security risks. There was a time when handheld devices were used as a perfect jumping-off point for hackers to break into the intranet, so it was correspondingly more difficult for businesses to manage and secure employees' personal devices.[2]

The importance of information security can never be overestimated. Every business is a part of the global, complex, and interdependent industrial chain, and it is the information that binds them together. From simple data to commercial confidential to intellectual property, the leakage of information can lead to damage to a company's reputation, financial losses, and even legal liability. Information plays a pivotal role in coordinating business relationships such as contracting and supply between companies.

As an increasing number of data is generated, stored, and analyzed, privacy issues will become more prominent, and new data protection requirements as well as improvements by legislatures and regulation administrations should be highlighted. Big data has considerable value, but also huge security risks, which is bound to cause huge losses to businesses and individuals if it is illegally used.[3]

3 Cybersecurity Law Builds the Wall of Privacy Protection for Citizens

On 21 August 2016, XU Yuyu, a girl in Linyi, Shandong who had just enrolled in a university, died of cardiac arrest suffering heartbreak and depression after being cheated out of her university fees by a fraudulent phone call. The cause of phone scams was the leakage of private information due to massive black market data trafficking. On 1 June 2017, the "Cybersecurity Law of the People's Republic of China" and related supporting regulations were formally implemented. The regulators took on the data chaos, and it was widely praised with dozens of data companies being investigated and tens of thousands of data interfaces being shut down. These companies being shut down had made data trading on the black market extremely prosperous, and even misleading the understanding of big data to whoever can trade data and whoever can trade as much private data as possible, becoming big data, among which there are even New Third Board listed companies valued at more than billions of Yuan. This act made companies operating informally to be kicked out of the market, and the underground data trading market totally shrink, also making investment institutions to be nervous. In fact, I have always had reservations about

---

[2] QI Dong, WU Yang, PENG Moxin. Facing up to the Challenges of Big Data on Information Security [J]. Keeping Secret, 2012 (8).

[3] DAI Yu. Promoting Government Decision-Making from "Informationalization" to "Big Data"— Interview with JIANG Qing, Director of Big Data Research Laboratory of the China Statistical Information Service Center [J]. South Reviews, 2015(19).

Data Security and Risk Management   119

such models as user data trading and big data credit, or that this is one of the pseudo-propositions of big data, which is a transaction detached from the data element itself. Instead, the core value of the development of big data is the mining value and result analysis based on the displaying laws of data, using data application research as an entry point to create a multi-scene model of data application.

The severe crackdown by public security departments and the official implementation of the cybersecurity law have made China's big data industry begin to formally react in terms of legality and rationality, the market space related to big data security has been further released, and the government and enterprises further increase their investment in big data security technology, data standardization, legal and trustworthy data sources, and innovation of big data products and services. China will also vigorously promote the dynamic operation of bilateral cross-border data and establish a coordination protection mechanism for data interaction between countries, thus people's enthusiasm to participate in the formulation of international standards and rules for cross-border data interaction will continue to increase.

4 The Pitfall of Big Data for Scientific Decision-Making

Now, there are two main pitfalls of big data for scientific decision-making: firstly, there is a risk that the predictive and analytical functions of big data will fail if the model algorithms are not scientific; secondly, there is the issue of data quality.

There are two ways to reduce the impact of false data: first, data sharing and openness, once achieved will force the data to be true under the supervision; second, big data has the characteristic that there will be other data sources to corroborate, therefore, when the amount of data large enough with enough data sources, the situation can be analyzed as close as much to the real situation based on correlated data, in which case a small amount of false data can hardly affect the overall results. Traditional research is a survey of the unknown, while big data is a record and aggregation of the known. When more and more data is collected intelligently, human intervention becomes less and less of a factor.

It should be clear that big data is now different from traditional sample statistical survey data. Big data places emphasis on full data acquisition, while sample surveys focus more on the scientific nature of sampling; big data places more emphasis on digging up the value behind the data, with more consideration of correlation studies, while traditional survey data rarely will pay attention to the relationship between A and B as it is already difficult to come up with the total, so the post-processing is basically the data classification and aggregation—not that it cannot, but is the traditional mindset—that request work has been done and they don't want to dig further or there is no such requirement for the job. If an organization or individual did a little more digging, it would be a different story.

There is indeed a shortage of talent in big data, and a very solid foundation of information technology, thinking, environment, and talent is still needed for the development of big data everywhere. What the government needs is not only IT people or statisticians but also complex talents with knowledge in management and communication, so that they can help leaders make better analyses and judgments in management decisions. Therefore, the government prefers to outsource because it does not have the relevant capabilities for the time being.

However, if they are completely outsourced, some companies will intercept the data. Of course, these companies might not necessarily give the data away, perhaps it is just for their own internal use. However, some departments or local governments cannot realize the problem that the value of data that you can't discover yourself, others have the ability to grasp it and study you with purpose. For example, a university professor doing a national project in Zhejiang got the demographic data of a certain region, but the data was brought back by an American student who analyzed it and found that this remote place had gathered a large number of doctors and post-docs, which is against common sense, and finally concluded that there was a top-secret military base in the area. It can be seen from this that the data can show many patterns and reveal a lot of information.

5 The Importance of Risk Management for Big Data

It is a fair world where the rewards are proportional to the risks. But the key to unlocking the value of big data is how to quantify risk with data. In the past, risk management and decision-making were mostly based on subjective empirical judgment, supplemented by data, which has led to a low level of risk management. And in the current environment of macroeconomic adjustment, industrial structure optimization, and intensified external competition, it is extremely important to quantify risks, improve the level of risk management, and thus enhance competitiveness with data.

The rise of big data and the advance of other technologies have dramatically reduced response times for governments and businesses, which make it available now to access and process the information they need for risk management rapidly that would have been unimaginable a decade ago.

According to a case study by Price Waterhouse Coopers (PwC), an Asian bank analyzed a complex set of $30 million in cash flow instruments under 50,000 different scenarios in less than 8 h. Whereas when big data and advanced processing power were not yet available, the same exercise could take weeks.

However, there are also obstacles to data sharing. Departments may defend their data for reasons such as confidentiality, fear of attracting too much scrutiny or losing control of certain work capital. Information silo is the enemy of big data risk management.

The increasing complexity of risk in different areas is driving the utilization of big data and attempts to control it. There is a consensus among economists and industry leaders that volatility will become a "new stability" in the next decade. Economic fluctuations, resource constraints, and political and social changes will all create an uncertain and unstable business environment for companies. In this context, the risk management of big data will become increasingly important.

Canadian Tire has conducted a groundbreaking investigation that links consumer behavior to credit risk. By analyzing in detail, the purchases made by consumers at multiple shops using credit cards issued by Canadian Tire, the company found that late deliveries and credit card defaults were all predictable. The approach was to study the types and brands of goods people purchase, as well as the types of bars they patronize. For example, data shows that consumers who purchase metal skull car accessories or modify large exhaust pipes are likely to end up not paying their bills. While 47% of customers who have spent money at Montreal's Sharx Pool Bar have defaulted on their payments four times in 12 months, making the bar the "riskiest bar" in Canada. The story of Canadian Tire illustrates a key problem with big data analysis: they can give you a complete picture. By bringing diverse data sets into the calculation, the understanding of risk can be improved thus reducing the risk.

In the era of big data, social data is proving to be an increasingly practical and straightforward risk management tool. Social media are effective early warning systems that can reflect changes in consumer sentiment, major macroeconomic risks, and even social and political risks. News of war and natural disasters may first come to light on social media such as Twitter, Facebook, and VK in Russia.[4]

Leaders can combine predictive analysis with techniques such as statistical modeling and data mining to make predictions about events, using them to assess potential threats. Big data companies have launched products to help customers conduct quickly experiments and develop rapid prototypes, allowing enterprises to try them out and even go on adventures before rolling them out on a large scale. The philosophy is based on the idea that learning from mistakes is an integral part of the development process. Leaders may therefore need to find ways to incorporate "learning from failure" into their processes, budgets, and capital allocations. This can be seen as a big step forward compared to the hindsight of risk analysis.

Given the sheer impact of big data, leaders are now at a crossroads. They can either do nothing and allow technological advances to commodify the skills they possess, leaving them in a diminishing position; or they can adapt to the new environment and increase their own influence and the value they can add to their organizations.

Big data means opportunities for all industries: taking on more strategic responsibilities and helping businesses to realize their future. Collecting and analyzing structured and unstructured data, modelling and examining information through big data technologies can provide managers with new and business-relevant services: making big data smaller and distilling information into incisive insights that can improve decision-making and achieve government and business transformation.

The advent of the era of big data has provided new opportunities and brought new challenges for all types of risk management. Big data belongs to the national foundation strategic resources, and there is an important impact and significance for departments to carry out overall data analysis and realize data-based scientific decision-making on their business. Only by fully developing and utilizing big

---

[4] ACCA. Big Data: Blessing or Disaster (Vol. 2) [J]. CFO World, 2014(5): 74–77.

data resources, continuously improving the scale, quality, and application level of departmental data, deeply exploring the potential value of departmental business data, testing and formulating work decisions with data, and using data to promote changes in risk management, can we continuously improve government governance capabilities.

In the future, when everyone is online, everything is connected, and the world is further sensed, an increasing number of data will be emerging. There are now over 2 billion smartphones in the world and over 600 million smartphones in China. The Internet of Vehicles, the Internet of Things, and the Internet of Machines have all become networks for data collection, circulation, and sedimentation, while the Chinese sensor market is expected to develop steadily and rapidly over the next five years. Currently, the Internet is also becoming an increasing capacity for collecting public data, thus mining valuable data from a large amount of public data will become a fundamental topic for a long time to come.

Big data processing technologies are becoming increasingly mature. Data search technology, speech recognition technology, image processing technology, face recognition technology, and unstructured text processing technology have all reached a relatively high level of intelligence and have basically met the needs of the application scenario. Visualization analysis is one of the important methods of big data analysis. Big data visualization analysis is based on the advantages of human cognitive ability for visual information, integrating the advantages of both humans and computers, with the help of human–computer interaction analysis methods and technologies, assisting people to grasp the information, knowledge, and laws behind big data more intuitively and efficiently. Big data visualization and analysis theory mainly include cognitive theory, information visualization theory, human–computer interaction, and user interface theory. Based on big data visualization technology, a series of big data visualization analysis and decision-making systems can be developed combined with industry applications, which can be widely used in aerospace, smart city, public safety, enterprise management, industrial control, and other fields.

Big Data, Big Opportunity, and Big Future

Qing Jiang

1 National Big Data Strategy Layout

What are the comments made by national leaders and leaderships in relevant departments regarding the importance of big data to a new round of technological revolution and industrial transformation, and the huge development opportunities that big data transformation holds? What is the top-level design?

On 5 March 2014, Premier LI Keqiang pointed out in his government work report at the second session of the 12th National People's Congress that a platform for entrepreneurship and innovation in new industries should be established to catch up with the advanced ones in new-generation mobile communications, integrated circuits, big data, advanced manufacturing, new energy, and new materials, and to lead the development of future industries. This is the first time the "big data" is being written into the government work report, which indicates that as a new industry, it will receive strong support from the national level.

During a discussion with Chinese and overseas delegates at the first World Internet Conference in Hangzhou on 20 November 2014, LI Keqiang told overseas Internet authorities attending the conference that "the direction of the Internet is not only e-commerce, but also cloud computing big data and the Internet of Things, where Chinese and foreign enterprises should have communications and cooperation, and I believe you will see a bigger market in China."

From finance, medical treatment, and trade to supporting e-government regulatory audits for small and micro enterprises, in LI Keqiang's view, the utilization of big data is by no means only a matter for enterprises, but also for government departments. On 25 July 2014, during a visit to the Inspur Group in Shandong, LI Keqiang asked the responsible persons of relevant departments to conduct "on-site work" by his side and requested them to use the concept of big data computing in the cloud

---

Q. Jiang (✉)
Zhongnan University of Economics and Law, Wuhan, Hubei, China
e-mail: 18712919290@163.com

© China Renmin University Press 2023
Q. Jiang (ed.), *Digital China: Big Data and Government Managerial Decision*,
https://doi.org/10.1007/978-981-19-9715-0_11

to organically dovetail with enterprise information technology platforms, as well as establish a unified integrated credit information platform for big data sharing. He said that "Whether it is to promote the government's streamline administration, decentralization, or to promote a new type of industrialization, urbanization, and agricultural modernization, all rely on big data computing in the cloud. Therefore, it should be the prevailing trend."[1]

On 14 October 2015, the Minister of Industry and Information Technology, MIAO Wei, published an article in the People's Daily entitled "Big Data, a Key Resource for Transforming the World", arguing that we are currently in the era of big data transformation. The mobile Internet, smart terminals, and new sensors are rapidly penetrating into every corner of the earth, as everyone has a terminal, everything can be sensed, everywhere can get online, and every moment can be linked, making it difficult to measure the growth rate of data as geometric or even explosive. Some institutions expect that by 2020, global data usage will reach approximately 44 ZB (1 ZB equals 10 trillion bytes), covering all areas of economic and social development. The resulting revolutionary impact will reshape the development of productivity, restructure the organization of production relations, improve industrial efficiency and management, and enhance the accuracy, efficiency, and predictability of government governance. There is no doubt that big data will create the next generation of Internet ecology, the next generation of innovation systems, the next generation of manufacturing, and the next generation of social governance structures. At the same time, big data will also change the mode of competition between countries. The reliance on data is rapidly increasing for countries, and the focus of international competition will shift from the competition for capital, land, and resources to the competition for big data, focusing on the scale and activity of data owned by a country, as well as the ability to analyze, dispose of, and use the data. Hence, digital sovereignty will become another major gaming area after border defense, sea defense, and air defense. Major countries have recognized the strategic significance of big data for a country, and whoever holds the initiative and dominance of data will win the future. The new round of great power competition is in large part about increasing influence and dominance over the world situation through big data.[2]

MIAO Wei's analysis above fully demonstrates the strategic significance of big data for the future of China. At present, China has the world's largest number of Internet users and mobile Internet users, the world's largest production base of electronic information products, the world's most growing information consumption market, and has cultivated a number of internationally competitive enterprises. The huge user base and complete economic system provide rich data resources, and the industrial internet will further stimulate the potential of the development of big data.

---

[1] Let Our People Benefit from the Reform—LI Keqiang's Trip to Shandong [J]. Openings, 2014(37).

[2] MIAO Wei. Big Data, a Key Resource for Transforming the World [N]. People's Daily, 2015-10-14.

As China's economic development enters the new normal, whether it is to maintain the medium-to-high speed of economic growth, promote the industry to the middle- and high-end level, or create a development environment of "mass entrepreneurship and innovation", big data will play an increasingly important role, and its fundamental, strategic, and pioneering position in economic and social development will become an increasing number of prominent.

On 1 July 2015, the General Office of the State Council issued "Several Decisions on the Utilization of Big Data to Strengthen Services and Supervision of Market Subjects" and established a timetable for action by various ministries and departments. On 19 August, the State Council's Executive Meeting adopted the "Outline of Action on Promoting the Development of Big Data", emphasizing the development and application of big data as a fundamental strategic resource, which should be in accordance with the strategic deployment of building manufacturing power and Internet power, strengthening the construction of information infrastructure, enhancing the support capacity of the information industry, building and improving the data-centered big data industry chain, promoting the opening and sharing of public data resources, and accelerating the development of core technologies, application models, and business models for collaborative innovation, thus turn big data into a new engine for upgrading the quality and efficiency of the economy under the new normal, as well as provide stronger support for economic development and social progress.

With regard to the opening up of public data resources, the State Council pointed out in its Action on Promoting the Development of Big Data that by the end of 2018, a unified open platform for national government data should be built, and reasonably and appropriately open to the public firstly in important areas such as credit, transportation, and healthcare. Currently, several provinces and cities in China have released their big data development strategies, and big data authority has been set up in Guangzhou, Shenyang, Chengdu, and others to ensure the promotion of the integration and opening of government data resources from the organizational structure.

On 8 December 2017, XI Jinping, general secretary of the Communist Party of China (CPC) Central Committee, stressed during the second group study session of the Political Bureau of the 19th CPC Central Committee that it was necessary to promote the national big data strategy, accelerate the improvement of digital infrastructure, promote the integration and open sharing of data resources, safeguard data security, accelerate the construction of digital China, and better serve the economic and social development of China as well as the improvement of people's livelihood. If big data was still widely discussed at the conceptual level in previous years, when some leadership did not even understand the important value of big data for economic and social development, this study has really touched on the importance of government departments for the applications of big data.

2 Interpretation of the Implementation of the National Big Data Strategy

The Outline of the 13th Five-Year Plan (2016–2020) for Economic and Social Development of the People's Republic of China, is prepared in accordance with the "Proposal of the Central Committee of the CPC on the Formulation of the 13th Five-Year Plan", which mainly sets out the strategic intent of the State, specifies the ambitious goals, major tasks, and significant initiatives for economic and social development, and serves as a guide for the behavior of market players. It is an important basis for the government to perform its duties and the common vision of the people of all nationalities. The Outline focuses on the implementation of the national big data strategy in Chap. 27.

Chapter 27 Implementation of the National Big Data Strategy

Take big data as a fundamental strategic resource, to fully implement actions to promote the development of big data, speed up the sharing and openness, as well as development and application of data resources, thus helping the industrial transformation and upgrading social governance innovation.

Section I. Accelerating Open Sharing of Government Data

Comprehensively promote the efficient collection and effective integration of big data in key areas, deepen the correlation analysis and integration of government data and social data, and improve the accuracy and effectiveness of macro regulation and control, market supervision, social governance, and public services. Relying on the unified government data sharing and exchange platform, accelerate the sharing of data resources across departments. Accelerate the construction of a national unified open platform for government data, and promote the interconnection and open sharing of government information systems and public data. Develop a catalog of government data sharing and opening, and promote the opening of data resources to society in accordance with the law. Coordinate the planning and construction of national big data platforms, data centers, and other infrastructure. Study and formulate laws and regulations on data opening and protection, and develop methods for managing government information resources.

Section II. Promoting the Healthy Development of Big Data Industry

Deepen the innovative application of big data in various industries, explore new modes of synergistic development with traditional industries, and accelerate the improvement of big data industry chain. Accelerate the research and development of key technologies in the fields of massive data collection, storage, cleaning, analysis and discovery, visualization, security, and privacy protection. Promote the development of big data software and hardware products. Improve

---

the public service support system and ecological system for big data industry, and strengthen the standard system and quality technology infrastructure.

Literally, it is clear that data can lead to greater leadership, insight, and process optimization for governments and businesses. Let's take a look at the expectations of the 13th Five-Year Plan for the national big data strategy and the pathway for big data to fully penetrate all areas of society by deconstructing some of the keywords in the Outline.

Keywords: Sharing and Openness, Development and Application

The big data industry chain is divided into three nodal areas: resource, technology, and research application.

Resource refers to the ownership of the data. In the era of the Internet+, with the spread of the Internet, any organization has the potential to own data resources, which can be its capital for continued realization. The expanding scope of information flow and sharing, the increasing symmetry of information, and the increasing value of data, make data resources as important to countries as energy. However, the lack and closure of data sources have been the main bottleneck restricting the application and development of big data. With the implementation of the national big data strategy, the government will continue to promote open data to ensure the accelerated development of the industry.

Technology refers to hardware and software and other technologies related to big data. There is a certain basis of China's software and hardware technology, and Internet-related industries have a place in the international arena. However, for the early development of big data in the past few years and even till now, hardware, fundamental software, information security, and so on are still the direct beneficiary. The development trend of big data is favored globally, especially for the smart city, in which the focus is the development of fundamental hardware and software. Over the past few years, smart cities were popularly developed and constructed everywhere. Although the primary stage of the smart city—the digital city—has not yet been completed, from the perspective of the correlation study of big data, it is an inevitable phenomenon that the digital city, data platform, intelligent life, and intelligent manufacturing will develop in parallel in the development process of the smart city in the long term.

Research application is the embodied part of the core value of big data, namely data mining analysis and application. Now, data mining analysis and application have not yet been effectively valued and promoted. However, from now until the future, the application of big data in government and business will mainly rely on analysis services and visualization, and in a market already dominated by technology, organizations based on data resources and application will also become the new development direction, while organizations with capabilities of outstanding innovation and better industrial integration will grow even faster.

The open sharing, development, and application of data is a new sign of a country's comprehensive competitiveness. It is foreseeable that the country will continue to promote the legislation for big data opening to accelerate the process of opening up government data.

In the practice of big data research and application, the National Bureau of Statistics is the earliest official national ministry to face up to the coming era of big data and leading in the application of big data in statistical business. The subordinate department China Statistical Information Service Center, was the first to set up a big data research service base in Xiamen, the first to build a big data research laboratory (Home Big Data) with enterprises, and the first to apply big data to industry decision-making reference, performance capacity enhancement, and livelihood services, the first to co-build a big data research institute for education based on industry management and the first to use big data to serve the leadership of Central Committee and the State Council for decision-making reference. According to data objects, based on business needs and problem orientation, in the China Statistical Information Service Center (Home Big Page), a big data research application platform that vertically integrates online and offline data was built, including statistical big data, public opinion (social opinion) big data, reputation big data, e-commerce big data, food safety big data, think tank big data, education big data, and etc.

Nevertheless, the application of big data research in China is still in its early stage. Although Internet data has been used more often by many start-up companies, this is still far from the end, and the research and application of data from government departments, enterprises, and individuals will be an important direction for the development of big data in the context of "Internet+".

Keywords: Industrial Transformation and Upgrading, Social Governance Innovation

The year 2016 is the first year of the 13th Five-Year Plan which is critical for China's economy as it is a key point of time entering a period of further adjustment and transformation. Promoting industrial transformation and upgrading is an important task during the 13th Five-Year Plan, and innovation in social governance is also a necessary path to enhance the country's overall competitiveness during the 13th Five-Year Plan.

Every leader in government and business should consider one important question: how to use big data to optimize and innovate their business. To understand "innovation" in terms of big data thinking, here's how I see it: as long as big data is integrated into the existing government performance and enterprise business ecosystem, it will continue to create new value and help organizations win in a competitive environment, which is exactly "industrial upgrading" and "governance innovation". I have been emphasizing on various occasions that "the Internet is about openness, while big data is about integration", which is to call on existing government departments and market players to look at the development and innovation of big data more from the perspective of integration, rather than being unique.

Every natural person should also consider how they can use big data to command their work and life. The government data openness proposed in the Outline will provide more policy support for social governance and entrepreneurship and innovation. The national security department has already been applying big data in the daily work of social governance, such as pattern summaries, character portraits, and trend predictions. The meteorological department has already done useful exploration in the field of meteorological data to serve people's livelihoods, and now is uploading its data collection to the cloud. It is hoped that this move will promote more social organizations and more entrepreneurs to cooperate and jointly explore the deep value of meteorological big data and open up China's meteorological big data industry.

I believe that building an application-level big data platform with independent technical strength, integrating domestic high-quality data service providers, government's livelihood data, and Internet data, providing comprehensive big data platform-based services like technology, building a business model based on big data analysis, digging deeper into the value of data resources, and driving industrial transformation and upgrading and social governance innovation through big data analysis will become new competitive advantages.

It is a fact that big data has comprehensively penetrated all areas of society and all aspects of life, so the application of big data in any industry should not be only focused on the fundamental research and development technology but must be developed in the direction of solving practical problems, which is respecting problem-oriented and demand-oriented, and finally result in validation.

Deepening Big Data Application in Industrial Sectors

With the in-depth implementation of the national big data strategy, the policy measures related to big data in various industries will be implemented one after another, the development environment of China's big data industry will be further optimized, and the demand for big data services in various areas of the economy and society will be further enhanced, and the scale of big data industry will continue to maintain a high growth trend.

The construction of national big data comprehensive experimental zones will promote the formation of special areas, and the construction of big data agglomerations and new industrialized big data demonstration bases that combine local industrial characteristics as well as application and development will also continue to be promoted. The National Development and Reform Commission, the Ministry of Ecology and Environment, the Ministry of Industry and Information Technology, the National Forestry and Grassland Administration, and the Ministry of Agriculture and Rural Affairs have all launched official opinions or programs on the development of big data in relevant industries. Big data policies are gradually extended to various industries and fields, further accelerating the promotion of big data applications. The National Development and Reform Commission, the Office of the Central Cyberspace Affairs Commission, and the Ministry of Industry and Information Technology have approved comprehensive experimental zones and national big data engineering laboratories, and local governments have also introduced planning and policies to promote the development of local big data industries by combining their own characteristics. Industrial big data will further promote the transformation and upgrading of traditional industries. With the gradually implemented "Guidelines of State Council on Deepening the Integration of Manufacturing and Internet Development", "Development Plan on Big Data Industry (2016–2020)", "Development Plan on Intelligent Manufacturing (2016–2020)", and other policies, China will further deepen the application and promotion of industrial big data in the industrial field, explore the establishment of the industrial big data center, implement the demonstration project of industrial big data application, and promote the traditional industrial transformation of intelligent manufacturing. The development of the application of big data in transportation, environmental protection, finance, and other industries, will provide extensive and powerful support for all aspects of social and economic development.

In terms of regional distribution, China's big data industry has formed five distinctive regions, namely the Beijing-Tianjin-Hebei region, the Yangtze River Delta, the Pearl River Delta, the western region, and the northeast region. Among them, the Beijing-Tianjin-Hebei region and the Pearl River Delta region have become the most innovative and dynamic regions, not only forming a healthy development structure of large and medium-sized enterprises but also emerging Beijing, Guangdong, Shanghai, Guizhou, Zhejiang, and Jiangsu six concentration provinces and cities of big data enterprises.

In terms of supply structure, the supply structure of China's big data market shows a quadrilateral shape, including digital technology enterprises such as Baidu, Tencent, and Alibaba; traditional IT vendors such as Huawei, Lenovo, Inspur, and yonyou; national teams of big data applications such as China Statistical Information Service Center (Social Opinion Research Center of the National Bureau of Statistics) and China Statistical Information Consulting Center; and big data related enterprises such as TRS, JUSFOUN, Sea Big Data related enterprises represented by Tops, Nine Times Square, VASTDATA, IZP Technologies, Home Page, and etc., covering the fields of data collection, data storage, data analysis, data visualization, and data security, and etc.

The Number of Innovation Actors in Big Data Is Increasing.

Not only are Internet companies, IT companies, and a large number of start-ups actively engaged in technological and product innovation, but many research institutions and social organizations are also engaged in these activities. Among them, technology enterprises are the main force of big data technological innovation, which has constantly expanded their industrial influence in the open-source environment, seizing the competitive ground, striving for the right to dominate technology development and standard setting, and striving to cultivate a new industrial ecology based on open source.

The Speed of Innovation and Industrialization in Big Data Is Significantly Increasing.

As new technologies in big data become more intelligent, and an increasing number of new products, the collaborative innovation mechanism between industry, academia, research, and application has taken shape, with the speed of technological productization and industrialization significantly increasing. The new technologies in big data cover the entire big data industry chain, from data collection and analysis to the availability of visualization. Investment subjects also began to gradually diversify. Leading Internet enterprises, traditional IT tycoons, cloud computing enterprises, data management, and big data analysis enterprises have all become investment targets in the big data sector. Google, Facebook, Apple, Amazon, IBM, Microsoft, and other enterprises are keen to acquire promising technology start-ups, with active investment and M&A activities involving large amounts of money. Some big data companies are both the target of investment and M&A, as well as the subject of investment. Companies with mature applications or core technologies are highly favored, as industrial applications remain a hot spot for investment and financing, with capital gathering towards companies that master industrial application products and services or companies with potential for industrial application development.

The Application of Big Data for Government Management and Public Services Is Getting Increasingly Widespread.

Through big data analysis, the government can better predict social and economic development trends, solve specific problems in urban management and public services, etc., relying on big data to continuously improve the scientific level of decision-making and refinement of management. On the one hand, government departments have a large amount of fundamental data resources at their disposal, and on the other hand, there is a strong demand for applications in areas such as urban management, security control, and administrative supervision. Big data has changed the traditional administrative mindset, from government information openness to data integration and sharing, through which big data is driving the government from "empirical governance" to "scientific governance". Accordingly, cooperation between government and enterprises has become one of the important driving forces for the development of big data. Major equipment manufacturers such as Huawei, telecom operators such as Unicom, and big data companies such as Oracle have strategically cooperated with local governments and built big data platforms to help in launching smart city strategies and promoting the implementation of big data applications.

Relevant Technological Innovations Will Accelerate the Development of Big Data Industry.

The integration and development of big data and information technology will give rise to new technologies, new modes, and new business models. The development of the Internet of Things will greatly enhance the acquisition capability of big data. Cloud computing and artificial intelligence is also being gradually integrated into big data analysis systems, and machine learning technology will effectively enhance the analysis capability of big data. The technical development of big data is increasingly connected to information technology such as the Internet of Things, cloud computing, and artificial intelligence. Many companies are already developing machine learning technologies, and the most advanced machine learning and artificial intelligence systems are moving beyond traditional rule-based algorithms, allowing computers to learn new things without explicit coding, creating systems that can understand, learn, predict, adapt and even operate autonomously, which will effectively enhance the capability of big data analysis. Advances in unstructured data processing technologies will greatly increase the data value. For information that cannot be represented by numbers or a uniform structure, such as text, images, sound, web pages, etc., which we call unstructured data. Unstructured data is not easy to process but contains a large amount of information. Technological innovations based on open source are continuously improving the level of unstructured data processing technology, which will greatly enrich the data value. The development of data visualization technology will strongly promote the popularity of big data applications. Data visualization technology allows not only the truth behind big data resources to be presented to the public, but also organizations to retrieve and process data when they are busy on business, providing users with a convenient and direct analysis display without any technical threshold restrictions. Visualization technology is bound to effectively drive the widespread adoption of Big Data across all industries.

Big Data in the "Belt and Road" Initiative

The "Belt and Road" initiative has pushed China to the second peak of development, and local governments can once again achieve a regional economic take-off by following the "Belt and Road" initiative. It is then the primary issue facing government leaders to adjust and upgrade the industrial structure.

In terms of the good opportunities brought by the "Belt and Road" to the regional economic take-off, local government leaders and related enterprises should make use of big data to conduct adequate industrial analysis and situation analysis, to clarify the current situation of local industries and identify core industries, marginal industries, and industries with potential, after which to combine with other regional data to find the core motive power of regional economic development.

Shaanxi Launches "Belt and Road Big Data Platform"

The International Center of National Bureau of Statistics, Xi'an University of Finance and Economics, China Statistical Information Service Center, Home Big Data, and China (Xi'an) Institute for Silk Road Research jointly built and launched the "Belt and Road Big Data Platform", which is the first internationalized professional database in China in response to the major initiative of building "Belt and Road". The project is the first international professional data bank in China to respond to the significant "Belt and Road" initiative, aiming at summarizing the basic situation, policies, progress, and useful experience of the construction of "Belt and Road" through real-time collection and analysis of related data, so as to formulate development strategies for local governments and promote enterprises, research institutions, individuals and so on to effective participant in "Belt and Road" and provide data support and policy consulting. The data platform relies on data analysis researchers with backgrounds in statistics, economics, public administration, mathematics, and computing, with the guidance of think-tank experts and scholar, regularly publishes the results of "Belt and Road" related indices, analyses, and researches various types of data, and forms various research reports to provide comprehensive information services for national, provincial, and municipal government as well as corporate users.

The research of big data applications in the think tank related to "Belt and Road" has been started. China (Xi'an) Institute for Silk Road Research | "Belt and Road" Big Data Research Center, jointly built by the International Center of National Bureau of Statistics, Shaanxi Provincial Ministry of Statistics, China Statistical Information Service Center, Xi'an University of Finance, and Economics, the China (Xi'an) Silk Road Research Institute, finished in 2015 the Research Report on the Current State of Industrial Development and the Positioning of "Belt and Road" Development in Shaanxi Province. This is an important reference report for provinces and cities to participate in the construction and development of the Belt and Road and is also the first think-tank research result based on technologies and methods of big data in China. Highlights from the Report are excerpted below.

Research Report on the Current State of Industrial Development and the Positioning of "Belt and Road" Development in Shaanxi Province 2015. Summary of Report Highlights

1. Remarkable Economic and Social Development in Recent Years in Shaanxi Province

In recent years, the Shaanxi provincial committee and the provincial government have been thoroughly implementing the Scientific Outlook on Development and never wavered in the determination of implementing the State strategy of regional development, accelerating the construction of "Building Rich, Harmonious, Beautiful Shaanxi" in the topic of scientific development, enriching the people and strengthening the province, in the way of transformation of the economic development mode, and in the purpose and basis of protecting and improving people's livelihood. The economic growth rate is among the top in the country, and the total economic volume is steadily rising. People's quality of life and the ecological environment is obviously improving.

2. The Profound Contemporary Background of the "Belt and Road" Initiative

In September and October 2013, during his visits to Central Asia and Southeast Asia, Chinese President XI Jinping proposed the great idea of jointly building the Silk Road Economic Belt and the 21st Century Maritime Silk Road, which received positive responses from the associated countries and great attention international-wide. When Premier LI Keqiang participated in the 2013 China-ASEAN Expo (CAEXPO), he stressed the importance of paving a maritime silk road for the Association of Southeast Asian Nations (ASEAN) and creating a strategic point to drive the development of the hinterland. The joint construction of the "Belt and Road" is a strategic concept proposed by the Chinese government in light of the profound changes in the international and regional situation, as well as the new situation and tasks facing China's development. It is meant to maintain the global free trade system and the open economic system, promoting cooperation among countries along the route to overcome the difficulties of the times and seek common development, which represents the profound contemporary background.[3]

---

[3] DONG Qingfeng. "Belt and Road" Initiative Brings Tanzania Development Opportunities [J]. New Business Weekly, 2015(7).

Big Data, Big Opportunity, and Big Future 135

3. Shaanxi Province as the "New Starting Point of the Silk Road Economic Belt"

The fundamental purpose of Shaanxi Province in implementing the "Belt and Road" initiative is to build a new starting point for the Silk Road Economic Belt and highland for inland reform and opening. In line with the "Belt and Road" initiative, Shaanxi will focus on building five centers for transport and logistics, science and technology innovation, industrial cooperation, cultural tourism, and financial cooperation, strengthening the development of the Guanzhong area, creating a highland for inland opening-up and leading the reform and open of the western provinces.

4. Measures Were Implemented to Build the "New Starting Point of the Silk Road Economic Belt"

Shaanxi Province has implemented a number of measures to build the "New Starting Point of the Silk Road Economic Belt", proposing 297 policies and measures in terms of either the "Belt and Road" or the "Silk Road Economic Belt", of which the forms include programs, opinions, notices, plans, exhibitions, fairs, conferences, seminars, symposiums, forums, the formation of new institutions, agencies or regional cooperation, and more; covering all aspects of the industry, trade, tourism, culture, administration, and so on. Each city in Shaanxi Province also has its own focus to promote the construction of the Silk Road Economic Belt.

5. In spite of Comprehensive Advantages, Shaanxi Province Is Facing Problems and Challenges

As Shaanxi Province owns a relatively comprehensive advantage among the five domestic provinces of the Silk Road Economic Belt, it is logically the leader in the construction of the Silk Road Economic Belt. However, in the construction of the "Silk Road Economic Belt", there are still disadvantages such as international economic downturn, increasing downward pressure on the domestic economy, and there's much more work yet to be done on the adjustment of the economic transformation structure. The shortcomings in the field of the financial sector, the influence of universities, the carrying capacity of resources, and air logistics need also urgent attention and remedy.

6. The Clear and Accurate Positioning of Shaanxi Province in the "Silk Road Economic Belt"

Culture and tourism, science and education, energy and chemicals, transportation and logistics, high-end manufacturing, trade and finance, and entrepreneurship and innovation are the main perceptions of Shaanxi Province in the construction of the "Silk Road Economic Belt", and have become the main support for Shaanxi Province to achieve the strategic objectives of the Silk Road Economic Belt.

The positioning of Shaanxi Province in the "Silk Road Economic Belt" took its own natural resource advantages as well as economic and social development characteristics, and also highlighted the advantages of Shaanxi Province in comparison with other provinces and regions in the "Silk Road Economic Belt"; it has clearly named the shortcomings and deficiencies in the current economic and social development of Shaanxi Province, and pointed out the core point in the construction of the "Silk Road Economic Belt", which is clear and accurate.

7. 2015 Silk Road Economic Belt Development Index of Shaanxi Province

The purpose of the Silk Road Economic Belt Development Index (SREBDI) is to measure the economic and social development of the provinces and regions in the Silk Road Economic Belt, as well as the stage of the Silk Road Economic Belt construction. The evaluation system of the Silk Road Economic Belt Development Index consists of a general index (SREBDI), a fundamental development index, and a positioning development index (with the threshold limit value at 1).

As shown in Table 1, the SREBDI of Shaanxi Province in 2015 was 0.396, indicating that the construction of the "Silk Road Economic Belt" in Shaanxi Province is in the initial stage; the fundamental development index and positioning development index were 0.426 and 0.373 respectively, indicating that Shaanxi Province has a strong foundation for its own development, and the construction of important areas to realize its positioning in the "Silk Road Economic Belt" has just begun.

Table 1 2015 SREBDI and sector indices in Shaanxi Province

Indices	2015
SREBDI	0.396
Fundamental development index	0.426
Positioning development index	0.373

2.1 Silk Road Cloud—Establishment of the "Belt and Road" Dynamic Database Operation Center

On 18 April 2015, under the media reports of Xinhua News Agency, People's Daily, Economic Daily, CCTV, and many other central media, the Silk Road Cloud—"Belt and Road" Dynamic Database Operation Center was announced to be launched in Shanghai VSATSH Co., Ltd. According to YU Jianguo, Director of the Research Institute of Information Technology of the Ministry of Civil Affairs and Chairman and General Manager of Shanghai VSATSH Co., Ltd, the database project is the first professional database in China to respond to the major initiative of jointly building "Belt and Road", which aims to collect and analyze data related to "Belt and Road" in real-time, summarize the basic situation, policies, progress, and useful experiences of the construction of the "Belt and Road", so as to provide data support and policy advice for local governments to formulate development strategies and promote effective participation of enterprises, research institutions, and individuals in the "Belt and Road".

According to the introduction, the database is divided into four sections based on its structure and functions: "Belt and Road" database, World View of "Belt and Road", Comprehensive Analysis, and Effectiveness Assessment, with the purpose of presenting comprehensive information on the "Belt and Road" and overseas condition. The first phase of Silk Road Cloud—"Belt and Road" Dynamic Database was intended to include several sub-databases such as Policy Communication, Economic and Trade Cooperation, Facility Connection, Information Service, Cultural Silk Road, Intangible Cultural Heritage of the Silk Road, Green Silk Road, Silk Road Tourism, and etc. The structure of the database can be divided into static data, dynamic data, image data, and statistical data according to the type of data. The static data will be updated in a timely manner by the project team, including supplementing and improving the descriptions of channels, sections, and other columns of the database; the dynamic information management includes the dynamic information gathered in real-time, reflecting the overall picture of the construction of "Belt and Road" through the real-time updating of multiple sub-databases.

It is known that the database will rely on data analysis researchers with professional backgrounds in communication, sociology, economics, public administration, mathematical statistics, etc., under the guidance of think tank experts and scholars, regularly release hotspots rankings, index systems, analysis reports, blue books, and more to the public, analyze and study all types of data to develop a variety of research reports to and provide users with comprehensive information service.

In addition, because of the undeniable importance of big data, the above-mentioned regions are not the only ones developing big data nationwide. According to incomplete statistics, more than 20 provinces and cities across the country have carried out plans for the construction of cloud computing centers and the deployment of cloud industries, including the "Propitious Clouds Project" in Beijing, the "China Cloud Valley" in Harbin, the "Sky Cloud Plan" in Guangzhou, the "Grassland Cloud Valley" in the Ordos, and the "Cloud Computing Headquarter" in Tianjin, all of these different plans are emphasizing the importance of big data. On this basis, big data applications regional-oriented will be developed rapidly in the future.

Conclusion—How to Adapt to the Era of Big Data

"The future has already arrived; it's just not yet evenly distributed rather than an illusion of the future. Big data has already all around us, but it has yet to find a suitable place in most industries." This is how YANG Bin, Vice President of Tsinghua University, sums up the current state of big data, which has become a fulcrum of our times.

Recognizing, understanding, appreciating, adapting, and integrating into the era of big data requires a comprehensive and objective approach to the characteristics of the times, training big data thinking, fostering a big data environment, and cultivating big data talent.

Big Data Thinking Needs to Be Trained Immediately. Wise leaders will move away from traditional ways of decision-making and management, and embrace the new mode of management decision-making in the era of big data. Thinking is the most important foundation for adapting to any era. Many people talk about big data now, but the truth is that only the top of the industry will own big data. As the world has irreversibly entered the era of digital technology, the development of all fields of China should be good at using big data thinking. Society should also be reconstructed with data thinking at the right time when no one should refuse big data thinking.

Though it is impossible for the individual to own big data, big data thinking can be used to manage and analyze daily data, and develop a habit of valuing data analysis for problem solving. To develop the habit of using big data thinking to solve problems, we need to be good at using external big data to guide our decisions. Some Internet companies' data can be considered as big data, such as BAT, which can query the popularity of a keyword through Baidu to decide their travel itinerary, or query Tmall's review data to decide whether they buy a product.

With big data thinking, we will naturally value the role of data and will manage it and analyze it in a more disciplined way. Leaders, in particular, will be ruthlessly eliminated by the times if they do not have the data thinking.

Both leadership and normal personnel should appreciate the power of data and develop their own big data thinking.

The Application Environment Needs to Be Fostered Faster. In the process of economic development, in the context of the central government's repeated emphasis

© China Renmin University Press 2023
Q. Jiang (ed.), *Digital China: Big Data and Government Managerial Decision*,
https://doi.org/10.1007/978-981-19-9715-0

on streamlining administration, delegating power, and transforming government functions, the government's role in macro-control and market supervision is essential. Government departments should put more effort into fostering an environment for industrial development, so as to foster a healthy environment for the integration of big data and industry as soon as possible.

The advent of the era of the Internet has brought China's economic and social rapidly development into line with international standards, and the future development of China is believed to speed up like China's high-speed rail. However, it is particularly important for the government to clearly position itself in the process of fostering an industrial environment, to create a fine environment for innovation and entrepreneurship and policy atmosphere, and to ensure the healthy development of enterprises with the integrity and legal system. The government should moreover provide information on various aspects for enterprises' reference, and can also provide think tank advice directly to enterprises at critical moments. The government should do a good service role through innovative regulation.

Big Data is the product of the market, an inevitable result of the development of the Internet, and the most market-oriented industry segment. The development of big data should truly require a healthy environment where the government builds up the infrastructure for the enterprise performing, which should be the future direction of the government.

Big Data Talents Need to Be Cultivated Urgently. Talent is the key to winning this new round of international competition in the era of Internet+ big data.

The era of big data requires not only composite talents of high-end management, research, and development, but also numerous personnel for infrastructure development, project implementation, and maintenance. China has opened general colleges and universities with cloud computing and big data-related majors. As of 1 March 2018, 283 colleges and universities have been approved to open undergraduate majors in big data, and a number of colleges and universities have started to recruit master and Ph.D. students majoring in cloud computing and big data.

Big data industrialization involves fields of big data science, big data technology, big data engineering, and big data application, with a huge need for talents. In the face of such a large demand gap, it is difficult to meet the development needs in a short period of time considering China's existing education level and efficiency of education institutions. The government should focus on building a good environment for the development of big data talents and speed up the implementation of forward-looking talent training programs. Although the construction of majors and research directions at the undergraduate and postgraduate education levels has started, the process of talent cultivation is relatively long. Big data applications are imminent, and the question at this stage is how to guarantee on-the-job training for the current needs of big data applications, which has become an issue that requires specific attention.

We should see and acknowledge the problems and challenges in the current development of big data in China, but also the new opportunities facing the development of our industry. As long as we seize the opportunity and pay attention to the three elements of training leaders in big data thinking, fostering a big data development environment, and cultivating composite big data talents, China will definitely usher in a more brilliant and competitive development prospect in the era of the Internet+ big data.

To the future, I would like to conclude this book with the views of MA Jiantang, Party secretary and Deputy Director of the Development Research Center of the State Council, on big data "There has never been a technological change like big data revolution, which, within a few short years, has transformed from an idea of a few scientists to a strategic practice of leading companies around the world, and then to a competitive strategy of major countries, forming a trend that cannot be ignored. and It is a historical trend that cannot be ignored and cannot be avoided. The Internet, the Internet of Things, cloud computing, smart city, and smart planet are enabling data to grow rapidly along Moore's Law, and a digital space parallel to physical space is being created. In the new digital world, data has become the most valuable factor of production, and countries and enterprises that respond to the trend and actively seek changes will rise to become the new leaders; organizations that are indifferent and stick to the rules will gradually be marginalized and lose the vitality and momentum of competition."

Yes, big data is opening up a new era, and reshaping a different economic society.

Printed in the United States
by Baker & Taylor Publisher Services